THE THERMOPHYSICAL PROPERTIES OF METALLIC LIQUIDS

The Thermophysical Properties of Metallic Liquids

Volume 2: Predictive Models

Takamichi Iida

Professor Emeritus of Materials Science and Processing
Osaka University, Japan

Roderick I.L. Guthrie

Macdonald Professor of Metallurgy at McGill University, and
Director and co-founder of the McGill Metals Processing Centre
McGill University, Montreal, Canada

OXFORD
UNIVERSITY PRESS

OXFORD
UNIVERSITY PRESS

Great Clarendon Street, Oxford, OX2 6DP,
United Kingdom

Oxford University Press is a department of the University of Oxford.
It furthers the University's objective of excellence in research, scholarship,
and education by publishing worldwide. Oxford is a registered trade mark of
Oxford University Press in the UK and in certain other countries

First Edition published in 2015
Impression: 1

Published in the United States of America by Oxford University Press
198 Madison Avenue, New York, NY 10016, United States of America

British Library Cataloguing in Publication Data
Data available

Library of Congress Control Number: 2015930577

ISBN 978–0–19–872984–6

Printed in Great Britain by
Clays Ltd, St Ives plc

Preface

High temperature materials processing, such as smelting, refining, casting, welding, crystal growth, melt spinning, zone melting, and spraying, is an extremely complex area in which many thermophysical properties of metallic liquids are intimately involved. Numerous studies have been made to unravel this complexity on the basis of thermodynamics, reaction kinetics, and hydrodynamic analyses.

Between the 1950s and 1980s, great progress has been made in the industrial technology and science of process metallurgy, particularly for iron- and steelmaking processes. The advances in the science of extractive metallurgy have been described in numerous books, such as *Physical Chemistry of Metals* by Darken and Gurry (1953) and *Physical Chemistry of Melts in Metallurgy* by Richardson (1974). Advances in science and technology relating to liquid metallic processing operations are covered by books such as *Physical Chemistry of High Temperature Technology* by Turkdogan (1980) and *Engineering in Process Metallurgy* by Guthrie (1989). These books cover a wide range of topics relating to the physical chemistry of melts (i.e. liquid metals, salts, and slags): structure, physical properties, thermodynamic properties, reaction kinetics, interfacial phenomena, related mass/heat transport phenomena, and attendant fluid flows. As such, they are of great significance for process metallurgists. However, the former books place an emphasis on chemical, and reaction kinetics, while the latter emphasize process kinetics in metallurgical systems, related mainly to iron- and steelmaking. For such macroscopic or continuum treatments, the values of thermophysical properties are generally taken as empirical constants.

The advent of semiconductors, metallic glasses, and other functional materials since the late 1950s, and ever-increasing demands for the manufacture of high quality metallic materials have broadened the scope of high temperature science and technology of metallic liquids. At the same time, metallurgy has expanded into materials science and materials engineering. As such, materials process science has come to be based not only on chemistry but also on physics. This is especially true for the efficient manufacture of high quality metallic materials, where a detailed knowledge of the thermophysical properties of metallic liquids in question, based on the theory of liquids, is required. This atomistic, or microscopic, approach to materials processing, in contrast to equilibrium chemical thermodynamics, and process kinetics, provides an essential underpinning to understanding and interpretation of various phenomena in liquid metallic processing operations. From this viewpoint, the present authors published a technical book *The Physical Properties of Liquid Metals* (1988). The purpose of that book was to introduce theoretical equations based on structure, semi-empirical (semi-theoretical) equations, empirical equations, together with methods of experimental measurement and experimental data, for the physical properties of liquid metals. This book was written for metallurgical and materials research workers.

Since the 1970s, there has been renewed interest in the thermophysical properties of almost all liquid metallic elements or simple metallic substances (i.e. liquid metals, semimetals, and semiconductors) following the advent of mathematical modelling techniques supported by powerful computers. Nowadays, computer simulation studies of materials processing operations, based on mathematical models, are widely used as a very useful tool for improving liquid metal and liquid metallic processing operations and product quality. Naturally, accurate and reliable data for the thermophysical properties of metallic liquids are indispensable, not only for the execution of computer simulations and for the development of mathematical models, but also for the direct solution of industrial high temperature processing operations. In particular, accurate data for almost all liquid metallic elements are first needed. To elaborate further on this matter, while multicomponent alloys are typically treated in liquid metallic processing operations, accurate and reliable data for the respective pure components of an alloy system are first needed as a starting point. In addition, we must go deeply into a study of an element's (i.e. simple substance) properties, in order to clearly understand the essence of a metallic liquid's thermophysical properties. For example, in the liquid state, the motions of atoms through the liquid are impeded by frictional forces set up by their nearest neighbours. Thus, the atomic transport coefficients, i.e. molecular viscosity and diffusivity, of metallic liquids are dominated by the frictional forces among atoms. Furthermore, the atomic transport coefficients cannot be entirely formulated in terms of thermodynamic properties alone, because the coefficients involve the movement of atoms. Owing to this, model theories, such as the hole theory or the rate process (activated state) theory, are unlikely to provide fruitful predictions of transport coefficients of metallic liquids.

A large number of research articles on the thermophysical properties of metallic liquids have been published in the last quarter century or so. Their results or main points have been organized in numerous review articles and books such as *Measurement and Estimation of Physical Properties of Metals at High Temperatures* by Mills in *Fundamentals of Metallurgy* edited by Seetharaman (2005). This book emphasizes the need for reliable data for the thermophysical properties of liquid materials (e.g. liquid metals) involved in high temperature processes. Mills states that thermophysical property data are beneficial in two ways: (1) in the direct solution of industrial problems, and (2) as input data for the mathematical modelling of processes. Unfortunately, the explanations for the various thermophysical properties of melts are condensed into 70 pages in that book, so details for each property are not described. (However, over 170 important articles are referred.)

As mentioned previously, there is an urgent need for reliable data for the thermophysical properties of metallic liquids. Indeed, accurate and reliable data for the thermophysical properties, even of liquid metallic elements, are still not necessarily plentiful. The absence of accurate data for metallic liquids is mainly due to the experimental difficulties in obtaining accurate values for these thermophysical properties at high temperatures. The measurements of the thermophysical properties of metallic liquids at high temperatures are time-consuming, expensive, and in some cases, impossible. Further, they require considerable experimental and metrological expertise. Thus, much effort has been directed towards the development of reliable models for accurate predictions of the thermophysical properties of metallic liquids. Even so, for many years, little progress has been made on reliable models for accurate predictions. Early in this century, a new approach to making such predictions was presented for several thermophysical properties of all liquid metallic elements.

As seen from this brief historical outline (presented above), describing scientific approaches to the manufacture of metallic materials, the purposes of materials process science in this area are (1) to provide a clear understanding of the structure and thermophysical properties of metallic liquids based on the liquid state physics, and (2) to develop models that can be used for accurate predictions of the thermophysical properties of metallic liquids, particularly for all, or almost all, liquid metallic elements. In the area of materials process science, both accuracy and universality are required of any model for predicting the thermophysical properties of metallic liquids. Other purposes of materials process science are (3) to provide appropriate evaluation of the thermophysical property data for metallic liquids, and (4) to act as a guide to creation of new types of materials. In order to achieve these four main purposes, materials process science must incorporate a blending in of knowledge in many subject area, including process metallurgy, materials science, condensed matter physics (particularly, liquid state physics), molecular physics, chemistry (particularly, theoretical chemistry, and inorganic chemistry), and finally metrology (particularly for high temperature experiments on the thermophysical properties of metallic liquids).

The present book is divided into two volumes. This book of two volumes is a completely revised version of the authors' previous book *The Physical Properties of Liquid Metals* (1988): the present book lays emphasis on both the *Fundamentals* (*Volume 1*) and on *Predictive Models* (*Volume 2*) for accurate predictions of the thermophysical properties of metallic liquids. The performances of models/equations for several thermophysical properties (e.g. sound velocity, surface tension, viscosity) of liquid metallic elements are quantitatively assessed by determining relative differences between the calculated and experimental property values.

Volume 1 is intended as an introductory text explaining the structure and thermophysical properties of metallic liquids for students of materials science and engineering, and also for research scientists and engineers who have an interest in liquid metal and liquid metallic processing. The authors have tried to give simple explanations. However, the level of its contents is necessarily relatively advanced, since knowledge of the various scientific fields, mentioned above, is blended into one.

It comprises nine chapters, i.e. Chapters 1–9. Several basic matters for understanding the thermophysical properties of metallic liquids and for developing reliable models to accurately predict their thermophysical properties, together with methods for assessment of models/equations, are briefly described in Chapter 1. Chapter 2 is an introduction to the structure of metallic liquids. Two fundamental quantities in the theory of liquids, i.e. the pair distribution function and the pair potential are described. In particular, a reasonably detailed description of the distribution function is presented, because an understanding of the thermophysical properties of liquids must be based on a fundamental understanding of a liquid's atomic arrangement. Chapter 3 is concerned with the density of metallic liquids. Although density, or number density (i.e. number of atoms per unit volume), is an indispensable and basic quantity, accurate data are not plentiful. A new model for the temperature dependence of liquid metallic element density (or volume expansivities) is introduced. Thermodynamic properties of a metallic liquid, i.e. evaporation enthalpy, vapour pressure, heat capacity, are outlined in Chapter 4. The velocity of sound in a liquid metallic element is described in Chapter 5. Dimensionless new common parameters for better predictions of several important thermophysical properties of metallic liquids can be extracted from their sound velocity data. The new common parameters give an indication of an atom's hardness or softness, as well as a piece of useful information about the structure of metallic liquids. The new parameters are also useful in discussions of anharmonic effects of atomic motions in metallic liquids. This theme is further developed in Chapter 6, dealing with surface tension, and Chapter 7, which covers the dynamic property, or transport property, viscosity, (also considered in volume 2). In Chapter 6, characteristic features of experimental data for metallic liquid surface tensions are identified. Reasons for large discrepancies among experimental data for metallic liquid viscosities are clarified in Chapter 7. In Chapters 6 and 7, the relationship between surface tension and viscosity for liquid metallic elements is discussed in some detail. Chapter 8 is concerned with diffusion in metallic liquids. Knowledge of diffusion is needed for many fields of engineering. However, even self-diffusivity data are extremely scarce; a predictive model for metallic liquid self-diffusivity, expressed in terms of well-known physical quantities, is presented. Metallic liquids, like solid metals, are characterized by high electrical and thermal conductivities. Fundamentals of electronic transport properties, i.e. electrical conductivity, or electrical resistivity, and thermal conductivity, of metallic liquids are discussed in Chapter 9, the final chapter. The essential points of methods for measuring density, surface tension, viscosity, diffusivity, electrical resistivity, and thermal conductivity are also described.

Volume 2 is designed for research scientists and engineers engaged in liquid metallic processing. In Volume 2, using as a basis the fundamental issues presented in Volume 1, we discuss models used for predicting accurate values of metallic liquid thermophysical properties.

It contains eight chapters, a glossary, and ten appendices. Essential points in building reliable models for accurate predictions of the thermophysical properties of liquid metallic elements are outlined in Chapter 10, from the standpoint of materials process science. Chapter 11 is devoted to the velocity of sound in liquid metallic elements. Useful dimensionless common parameters (or dimensionless numbers), which characterize the metallic liquid state, can be revealed through data for the velocity of sound. These common parameters allow for better predictions of several thermophysical properties of liquid metallic elements. Models, in terms of the common parameters, are discussed for the volume expansivity, evaporation enthalpy, surface tension, viscosity, and self-diffusivity of liquid metallic elements, in Chapters 12–16. The performances of the various models are evaluated by comparing them against experimental values (provided experimental data are available). Predicted, or calculated, data for sound velocity, volume expansivity, evaporation enthalpy, surface tension, viscosity, and self-diffusivity of liquid metallic elements are given in Chapters 11–16. In Chapters 10, 11, and 13–16, atomic periodicity in values of each thermophysical property, discussed in this book, is illustrated for a large number of liquid metallic elements. In Chapter 17, the final chapter, a large number of experimental data for the physical quantities and the thermophysical properties of liquid metallic elements are compiled, although we emphasize that this book is not primarily a reference data book.

Prior to the Appendices, a glossary and/or supplementary explanations are provided. The periodic table contains many pieces of useful information about the thermophysical properties of elements; one form of periodic table is given in Appendix 1. Appendices 2 and 3 provide numerical expressions for determining the minimum values of relative standard deviation. SI units, unit conversions, fundamental physical constants in SI units, and the Greek alphabet, are given in Appendices 4–7. Occam's razor is an essential guide to the development of any model for accurate predictions of the properties of materials; Appendix 8 cites Occam's razor. In Appendices 9 and 10, calculated values of isothermal compressibility, structure factor, and the ratio of heat capacity of some liquid metallic elements at their melting point temperatures, are all listed.

Over the last half century, a huge number of research articles and review articles on the thermophysical properties of metallic liquids have been reported. Even so, our present knowledge of the thermophysical properties of metallic liquids is still lacking from the materials process science and engineering points of view. Accurate and reliable data for the thermophysical properties of metallic liquids are still not plentiful. Systematic investigations based on theory and experiment are greatly needed from the standpoint of materials process science and engineering. As such, we hope that this two volume book will not only be used for obtaining the relevant constants for the properties of specific liquid metallic elements, but will also help the user recognize the continuum between the microscopic and macroscopic approaches to liquid metallic processing operations. For the future, we hope this book will help in enabling great progress in materials process science.

Takamichi Iida
Roderick I.L. Guthrie
August, 2015

Acknowledgements

We wish to express our gratitude towards the offices of the McGill Metals Processing Centre for infrastructure support for this long running endeavour.

We thank Professor S. Iitaka of Gakushuin University, and Professor Emeritus H. Ishigaki of Waseda University for their helpful comments on the mathematical treatment given in Appendices 2 and 3. Similarly, we wish to thank Dr I. Yamauchi for his patient and expert work in producing the final typescript of the present, two-volume text book.

The first author sincerely thanks his wife, the late Sugaho, for her devoted assistance in typing the draft manuscripts and numerical calculations. He is also grateful to his daughter, Eri, for her continued cooperation.

Finally, we would like to express our sincere gratitude towards the Canadian Immigration Authorities and the Natural Science and Engineering Research Council of Canada for their support in allowing this extended period of collaboration in Canada.

Contents

Volume 2 Predictive Models

Volume 1 Fundamentals

Principal Symbols

Numbers in parentheses refer to equations.

Capital Italic

C	correction factor (10.5)
C_A	Andrade coefficient
C_P, C_V	heat capacity at constant pressure, at constant volume
C_T	constant
C_0	$\equiv 3.771 \times 10^{-2}\, k_0^{-1}$
D	self-diffusivity
D_{HS}	self-diffusivity in the hard-sphere fluid
E_C^0	molar cohesive energy at 0 K at 1 atm (101.325 kPa)
H_μ, H_D	constants or parameters (apparent activation energy for viscous flow, for diffusion)
$\Delta_l^g H_b$	enthalpy of evaporation at the boiling point temperature
M	molar mass
N	number of samples; number of atoms
N_A	Avogadro constant
Q	parameter (10.3), $Q = C$
R	molar gas constant
S, $S(N)$	relative standard deviation
$\Delta_l^g S_b$	entropy of evaporation at the boiling point temperature
$S(0)$	structure factor ($\lim_{Q \to 0} S(Q) = S(0)$, $Q = 4\pi \sin\theta / \lambda$, λ : wave length)
T	absolute temperature
U	sound velocity
V	molar volume; volume

Lower Case Italic

a, b	constants
e	electron charge
k	Boltzmann constant
k_s	parameter (14.2), $k_s \equiv \gamma_0 \xi_T^{1/2}$
k_v	parameter (15.6), $k_v = 0.1019 / k_0 \xi_T^{1/2}$ (15.8)
k_0	dimensionless numerical factor

m	atomic mass
n	repulsive exponent; principal quantum number
q_i	numerical factor

Greek

α	volume expansivity
$\underline{\alpha}$	parameter related to the distance over which the interatomic force extends
β	correction factor (for Lindemann's equation)
γ	surface tension
$\gamma_G, \gamma_{G,E}, \gamma_{G,T}$	Grüneisen constants
γ_h	$\equiv C_P / C_V$
γ_0	$\equiv k_0 C_A$
$\Delta, \Delta(N)$	global delta (1.21)
δ_i	relative difference between experimental and calculated values for χ_i
ε	parameter (5.27)
η	packing fraction
κ_S, κ_T	isentropic, isothermal, compressibility
Λ	temperature dependence of the density of liquid metallic elements
λ	thermal conductivity
μ	viscosity
ν	atomic frequency
ξ_E, ξ_T, ξ	dimensionless common parameters ($\xi \equiv \xi_T / \xi_E$)
ρ	density
ρ_e	electrical resistivity
σ	hard-sphere diameter
σ_e	electrical conductivity
Φ	total potential energy
χ_i	thermophysical property (e.g. sound velocity, surface tension, viscosity, volume expansivity, etc.)

Subscripts

b	at boiling point
m	at melting point

Selected General References

1. D. Tabor, *Gases, Liquids and Solids and Other States of Matter*, 3rd ed., Cambridge University Press, 1991.
2. G.S. Rushbrooke, *Statistical Mechanics*, Oxford University Press, 1949.
3. C. Kittel, *Introduction to Solid State Physics*, 8th ed., John Wiley & Sons, 2005. (See also 2nd ed., 1956; 7th ed., 1996.)
4. E.A. Moelwyn-Hughes, *Physical Chemistry*, Pergamon Press, 1961.
5. D.H. Trevena, *The Liquid Phase*, Wykeham Publications, 1975.
6. M.A. White, *Properties of Materials*, Oxford University Press, 1999.
7. W. Benenson, J.W. Harris, H. Stocker, and H. Lutz (eds.), *Handbook of Physics*, Springer-Verlag, 2002.
8. A. Stwertka, *A Guide to the Elements*, 2nd ed., Oxford University Press, 2002.

10

A Guide to Predictive Models for Metallic Liquid Thermophysical Properties

10.1 Introduction

The main purpose of materials science and engineering is to make the best use of all the elements in the periodic table; in other words, it is necessary to bring out their specific strengths, or their unique properties. For this purpose, in the field of high temperature materials processing, such as smelting, refining, casting, welding, crystal growth, not only accurate and reliable data for the thermophysical properties of all metallic liquids is needed, but also a clear understanding of the essence of their thermophysical properties, based on these data, is indispensable. In addition, while multicomponent alloys are typically treated in liquid materials processing operations, accurate and reliable property data for the respective pure components (i.e. liquid metallic elements) of an alloy system are first needed. Nevertheless, accurate experimental determinations of the thermophysical properties of all metallic elements are actually impossible. Over the last two centuries or so, various models have been proposed for the thermophysical properties of simple liquids, or liquid (metallic) elements, using many different angles and perspectives. Unfortunately, broadly speaking, none of these models are satisfactory from the viewpoint of materials process science.

We will now consider essential points in building reliable models for accurate predictions of the thermophysical properties of liquid metallic elements for which experimental data are lacking.

10.2 Essential Conditions of Predictive Models

10.2.1 Accuracy and Universality

In liquid metallic processing science, accurate data for the thermophysical properties of all liquid metallic elements are essential for obtaining a detailed knowledge of their characteristic, unique properties. Thus, both *accuracy* and *universality* are required of

The Thermophysical Properties of Metallic Liquids: Volume 2 – Predictive Models. First Edition.
Takamichi Iida and Roderick I. L. Guthrie. © Takamichi Iida and Roderick I. L. Guthrie 2015.
Published in 2015 by Oxford University Press.

any model for predicting thermophysical properties of metallic liquids. For this, the model should be expressed in terms of well-known universal quantities whose values are, or are nearly, established (e.g. molar mass (or atomic mass), atomic number, melting point, density (or molar volume)). Incidentally, in building up a model, an essential guide to the scientific method is *Occam's Razor* (see Appendix 8), and this must be uppermost in our mind. According to *Occam's Razor*, a model or theory should be no more complicated than necessary, i.e. the simplest explanation is usually the best, in terms of developing the science of a subject.

10.2.2 Quantitative Assessment of Models

The performance of models or equations, if we can, needs to be evaluated quantitatively; we have already used three parameters (i.e. δ_i, $\Delta(N)$, and $S(N)$) to assess our models (see Section 1.5). The three parameters, δ_i, a global $\Delta(N)$, and a relative standard deviation, $S(N)$, are defined, respectively, as:

$$\delta_i = \frac{(\chi_i)_{\text{exp}} - (\chi_i)_{\text{cal}}}{(\chi_i)_{\text{cal}}} \times 100, \ (\%) \tag{1.20}$$

where $(\chi_i)_{\text{exp}}$ and $(\chi_i)_{\text{cal}}$ are the experimental and calculated values for the thermophysical property χ of a liquid metallic element i, respectively. The parameter δ_i indicates the relative differences between experimental and calculated values for a metallic liquid's thermophysical property.

$$\Delta(N) = \frac{1}{N}\sum_{i=1}^{N} |\delta_i|, \quad (\%) \tag{1.21}$$

where N is the number of samples.

$$S(N) = \frac{1}{100}\sqrt{\overline{\delta_i^2}} = \sqrt{\frac{1}{N}\sum_{i=1}^{N}\left\{\frac{(\chi_i)_{\text{exp}} - (\chi_i)_{\text{cal}}}{(\chi_i)_{\text{cal}}}\right\}^2} \tag{1.22}$$

It is of great importance that the performance of a model is assessed by comparing (i) accurate and reliable experimental values (ii) for a large number of samples. We now give examples of these problems.

(i) Table 10.1 compares the performances of the Schytil model [1] (see Subsection 6.3.2) for the surface tensions of high melting point liquid metals ($T_m > 2000$ K) at their melting temperatures. In Table 10.1, on the basis of the following equations, Eqs. (10.1) and (10.2), calculated values $(\gamma_m)_{\text{cal}}$ for their surface tensions are compared against selected experimental values $(\gamma_m)_{\text{exp}}$ picked from references [2–4] by each of research groups (a) and (b), respectively.

As can be seen, different experimental values are used in groups (a) and (b) for the same metals, except for thorium.

$$\gamma_m = 4.495 \times 10^{-8} \frac{RT_m}{V_m^{2/3}}, \quad \text{for group (a)} \tag{10.1}$$

$$\gamma_m = 4.416 \times 10^{-8} \frac{RT_m}{V_m^{2/3}}, \quad \text{for group (b)} \tag{10.2}$$

where the numerical factors of 4.495×10^{-8} mol$^{1/3}$ and 4.416×10^{-8} mol$^{1/3}$ were determined so as to give the minimum S values (see Appendix 2) for the 15 metals shown in Table 10.1. The experimental values for their surface

Table 10.1 *Perfomance of the Schytil model for the surface tensions of high melting point liquid metals at their melting point temperatures, in which different experimental values, $(\gamma_m)_{exp}$, are used for the same metals, except for thorium.*

Group (a)

Metal		$(\gamma_m)_{exp}$ N m^{-1}	$(\gamma_m)_{cal}$ / N m^{-1} Eq. (10.1)	δ_i %
Chromium	Cr	1.710	1.992	−14.2
Hafnium	Hf	1.490	1.469	1.4
Iridium	Ir	2.230	2.248	−0.8
Molybdenum	Mo	2.290	2.286	0.2
Niobium	Nb	1.937	1.972	−1.8
Osmium	Os	2.500	2.762	−9.5
Platinum	Pt	1.673	1.610	3.9
Rhenium	Re	2.710	2.804	−3.4
Rhodium	Rh	1.915	1.860	3.0
Ruthenium	Ru	2.260	2.208	2.4
Tantalum	Ta	2.360	2.332	1.2
Thorium	Th	1.006†	0.960	4.8
Tungsten	W	2.500	2.898	−13.7
Vanadium	V	1.950	1.819	7.2
Zirconium	Zr	1.400	1.252	11.8
			Δ(15) %	5.3
			S(15)	0.070

continued

Table 10.1 *(continued)*

Group (b)

Metal		$(\gamma_m)_{exp}$ N m^{-1}	$(\gamma_m)_{cal}$ / N m^{-1} Eq. (10.2)	δ_i %
Chromium	Cr	1.630	1.957	−16.7
Hafnium	Hf	1.614	1.443	11.9
Iridium	Ir	2.140	2.208	−3.1
Molybdenum	Mo	1.915	2.246	−14.7
Niobium	Nb	1.839	1.937	−5.1
Osmium	Os	2.400	2.714	−11.6
Platinum	Pt	1.800	1.582	13.8
Rhenium	Re	2.610	2.754	−5.2
Rhodium	Rh	2.000	1.827	9.5
Ruthenium	Ru	2.180	2.169	0.5
Tantalum	Ta	2.016	2.291	−12.0
Thorium	Th	1.006†	0.943	6.7
Tungsten	W	2.200	2.847	−22.7
Vanadium	V	1.855	1.787	3.8
Zirconium	Zr	1.500	1.230	22.0
			Δ(15) %	10.6
			S(15)	0.124

† Only one datum point is reported [2, 3].

tensions, reported in [2–4], vary approximately ± 2 to ± 10 per cent around the mean; these discrepancies will probably be about average compared with those for other thermophysical properties and physical quantities.

As is clear from Table 10.1, large differences exist between the Δ and *S* values of the (a) and (b) groups (cf. Table 14.3). The results of Table 10.1 suggest the following: if we may be allowed an exaggeration, then it is possible to propose any model, because desirable experimental data that fit the model can easily be selected. In other words, there are therefore quite a few models (or theories) that coincide completely with inaccurate, or wrong, experimental data. Thus, accurate and reliable experimental data are indispensable for surface tensions (or more generally, for any thermophysical property) of metallic liquids in order to check the true capability of any model and to then devote our professional time to subsequent studies. Otherwise, castles in the air will be built. Furthermore, we repeat here that one of the most important purposes of materials process science and engineering is to precisely evaluate values for thermophysical properties of metallic liquids.

(ii) Figure 10.1 shows a plot of experimental viscosity data against $(MT_{\mathrm{m}})^{1/2} / V_{\mathrm{m}}^{2/3}$ (i.e. an Andrade relationship [5]) for 26 liquid metallic elements. It shows a good linear correlation between the two variables. The experimental viscosity values plotted in Figure 10.1 are listed in Table 17.9. Tables 10.2(a) and (b) compare the experimentally derived viscosity values with those calculated from the Andrade formula. As is obvious from Table 10.2(a), the Andrade formula provides $\Delta(26)$ and $S(26)$ values of 7.3 per cent and 0.080, respectively: the agreement obtained between calculation and experiment for these metallic elements is very good. Table 10.2(b), however, shows that agreement with experimental values for gallium, iron, plutonium, titanium, tungsten, and yttrium is only fair. For 48 liquid metallic elements, the Andrade formula gives $\Delta(48)$ and $S(48)$ values of 14.9 per cent and 0.182, respectively (see Section 7.6). The performance of a model, if possible, should be assessed for both many and various kinds of samples.

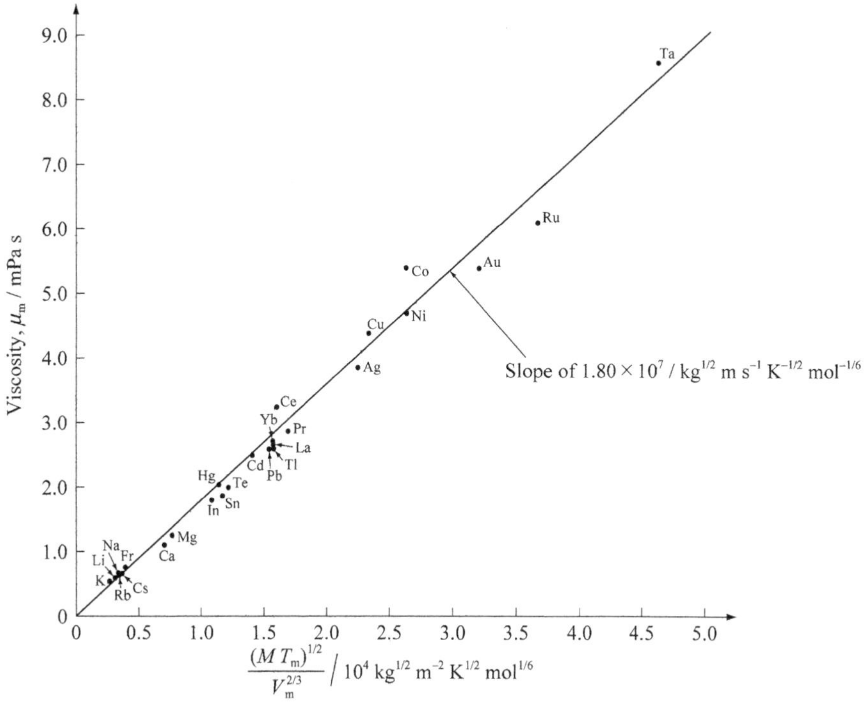

Figure 10.1 *Viscosity of various liquid metallic elements at their melting points vs. $(MT_m)^{1/2} V_m^{-2/3}$, i.e. the Andrade relationship.*

Table 10.2 *Comparison of experimental values for the melting point viscosity of liquid metallic elements with those calculated from the Andrade formula, together with δ_i, Δ, and S values.*

(a)

Element		Viscosity μ_m / mPa s		δ_i
		Experimental	Calculated	%
Cadmium	Cd	2.50	2.53	−1.2
Caesium	Cs	0.686	0.657	4.4
Calcium	Ca	1.10	1.26	−12.7
Cerium	Ce	3.25	2.90	12.1
Cobalt	Co	5.4	4.76	13
Copper	Cu	4.38	4.20	4.3
Francium	Fr	0.765	0.707	8.2
Gold	Au	5.37	5.79	−7.3
Indium	In	1.79	1.97	−9.1
Lanthanum	La	2.66	2.84	−6.3
Lead	Pb	2.59	2.78	−6.8
Lithium	Li	0.602	0.567	6.2
Magnesium	Mg	1.25	1.38	−9.4
Mercury	Hg	2.048	2.06	−0.6
Nickel	Ni	4.7	4.76	−1.3
Potassium	K	0.537	0.499	7.6
Praseodymium	Pr	2.85	3.05	−6.6
Rubidium	Rb	0.674	0.623	8.2
Ruthenium	Ru	6.1	6.62	−7.9
Silver	Ag	3.88	4.05	−4.2
Sodium	Na	0.695	0.617	12.6
Tantalum	Ta	8.6	8.33	3.2
Tellurium	Te	2.0†	2.20	−9.1
Thallium	Tl	2.59	2.85	−9.1
Tin	Sn	1.87	2.11	−11.4
Ytterbium	Yb	2.67	2.83	−5.7
			$\Delta(26)$ %	7.3
			$S(26)$	0.080

† Mean value (see Table 17.9(a)).

Table 10.2 *(continued)*

(b)

Element		Viscosity μ_m / mPa s		δ_i %
		Experimental	Calculated	
Gallium	Ga	1.95	1.63	19.6
Iron	Fe	5.6	4.55	23
Plutonium	Pu	5.95	4.49	32.5
Titanium	Ti	4.42	3.39	30.4
Tungsten	W	7.0	9.85	–29
Yttrium	Y	3.6	2.96	22
			Δ(6) %	26.1
			S (6)	0.265

In liquid metallic processing science, predictions play a vital role; a model is generally considered valid if it can be used to accurately predict values for the thermophysical properties of metallic liquids for those not yet available. Nevertheless, accurate predictions are extremely difficult.[1] It may be fairly easy to build models that fit experimentally obtained data for thermophysical properties of metallic liquids, or models may be adjusted so as to fit presently available experimental values, but this does not necessarily mean the models can predict accurate values for the thermophysical properties of metallic liquids that have not yet been measured experimentally, although good agreement with all existing- experimental data is an essential condition of a predictive model.

10.3 Improvements on Well-Known, Representative Models for Liquid Thermophysical Properties

In Volume 1, we saw that all the models (or equations) shown in Table 1.2, though they are well-known, representative expressions, have weak points: broadly speaking, the common weakness of these models is that the agreement obtained between calculation and experiment is not necessarily good, and, as such, the accuracy of these models is not satisfactory from the standpoint of materials process science. In order to accurately predict the thermophysical properties of liquid metallic elements, Iida et al. have introduced two new dimensionless parameters, $\xi_T^{1/2}$ and $\xi_E^{1/2}$, deduced from data for their respective sound velocities (see Chapter 5). Table 10.3 summarizes the results obtained in Volume 1, i.e. the performances of representative models, a modified

[1] In this book, 'to accurately predict values' or 'accurate predictions' indicates that predicted values fall, or almost fall, within the range of uncertainties of experimental measurements likely to be made in the future.

Table 10.3 *Representative models (or equations) for the thermophysical properties of solid and liquid metallic elements, and their performances together with those of the modified models proposed in Volume 1.*

Model in SI units	Researcher	Published year	Performance / Δ(N)%, S(N)		Note or comment
			Original, $f(\xi^{1/2})$	or modified	
Troution's rule, liquid→gas	Trouton [6]	1883	$\Delta(62)$ 13.5	–	(a)
$\Delta_l^g S_b = \frac{\Delta_l^g H_b}{T_b} \approx \text{constant}$			$S(62)$ 0.161		
Lindemann's equation, solid	Lindemann [7]	1910	–	–	(b)
$\nu_m = 8.9 \times 10^8 \left(\frac{T_m}{M V_m^{2/3}} \right)^{1/2}$			–	–	
Steinberg model, liquid	Steinberg [8]	1974	$\Delta(16)$ 17.9	$\Delta(16)$ 13.7	(c)
$\alpha_m = \frac{0.23}{T_b - 0.23 T_m}$			$S(16)$ 0.229	$S(16)$ 0.167	
Gitis–Mikhailov model, liquid	Gitis and Mikhailov [9]	1968	$\Delta(39)$ 24.7	–	(d)
$U = \left(\frac{2E_C}{M} \right)^{1/2}$			$S(39)$ 0.291	–	
Einstein model, sold	Einstein [10]	1911			
$U = q_1 \nu V^{1/3}$					
Einstein–Lindemann model, solid			$\Delta(41)$ 24.8	$\Delta(41)$ 14.5	(e)
$U_m = q_2 \left(\frac{R T_m}{M} \right)^{1/2}$			$S(41)$ 0.293	$S(41)$ 0.170	

Skapski model, liquid	Shapski [11]	1948	Δ(39) 28.6	Δ(39) 19.8	(f)
$\gamma_m = q_3 \frac{\Delta_l^g H_b}{V_m^{2/3}}$			S (39) 0.373	S (39) 0.241	
Schytil model, liquid	Schytil [1]	1949	Δ(41) 35.1	Δ(41) 19.8	(g)
$\gamma_m = q_4 \frac{RT_m}{V_m^{1/3}}$			S (41) 0.436	S (41) 0..241	
Fowler–Born–Green relation, liquid	Fowler [12]	1937			(h)
$\mu = \frac{16}{15}\left(\frac{m}{kT}\right)^{1/2}\gamma$	Born and Green [13]	1947	Δ(37) 32.7	Δ(37) 22.2	
			S (37) 0.400	S (37) 0.269	
Andrade formula, liquid	Andrade [5]	1934	Δ(48) 14.9	–	(i)
$\mu_m = 1.80 \times 10^{-7} \frac{(MT_m)^{1/2}}{V_m^{2/3}}$			S (48) 0.182	–	
Modified Stokes–Einstein formula, liquid	Eyring and Ree [14]	1961	Δ(18) 7.8	–	(j)
$D = \frac{kT}{\varsigma(V/N_A)^{1/3}\mu}$			S (18) 0.102	–	
Wiedemann–Frantz–Lorenz law, solid	Lorenz [15]	1872	–	–	(k)
$\frac{\lambda}{\sigma_e T} = \frac{\pi^2 k^2}{3e^2} = 2.45 \times 10^{-8}$			–	–	

continued

Table 10.3 *(continued)*

List of symbols

$\Delta_l^g S_b$	molar evaporation entropy at T_b	U	sound velocity
T	absolute temperature	R	molar gas constant
M	molar mass	μ	viscosity
α	volume expansivity	k	Boltzmann constant
E_c	molar cohesive energy	ζ	constant (5 to 6)
γ	surface tension	N_A	Avogadro constant
m	atomic mass	σ_e	electrical conductivity
D	self-diffusivity	q_i	numerical factor ($i = 1$ to 4)
$\Delta_l^g H_b$	molar evaporation enthalpy at T_b	λ	thermal conductivity
ν	mean atomic frequency	e	electron charge
V	molar volume		

Subscripts

m at melting point
b at boiling point

Notes or comments

(a) The following equations have been proposed for the molar evaporation enthalpy of liquid metallic elements.

i) A relationship between molar evaporation enthalpy $\Delta_l^g H_b$ at the boiling point and molar cohesive energy E_C^0 at 0 K at 1 atm:

$$\Delta_g^l H_b = 0.889 E_C^0 \tag{4.6}$$

Eq. (4.6) performs excellently with $\Delta(59)$ and $S(59)$ values of 4.8 per cent and 0.063, respectively.

ii)

$$\Delta_g^l H_b = 352 \xi T_m \tag{4.10}$$

The value of ξ ($\equiv \xi_T/\xi_E$) vary periodically with atomic number. On the basis of Eq. (4.10), we can predict values for the evaporation entropy $\Delta_l^g H_b$ *of* liquid metallic elements using their melting point temperature data.

(b) Lindemann's equation for the mean atomic frequency of solids at their melting points has been extended to that of liquid metallic elements at their melting points:

$$\nu = 6.8 \times 10^{11} \left(\frac{\gamma_m}{M} \right)^{1/2} \tag{6.36}$$

(c) The agreement between experimental and calculated values for the zinc group metals, using the Steinberg model (i.e. $\delta_{Cd} = -40.4$, $\delta_{Hg} = -55.4$, and $\delta_{Zn} = -33$ per cent) is unsatisfactory.
The model, represented in terms of $\xi_E^{1/2}$ gives Δ (16) and S(16) values of 13.7 per cent and 0.167, respectively (see Table 3.4).

(d) A dimensionless common parameter $\xi_E^{1/2}$ can be extracted from the Gitis–Mikhailov model.

(e) A dimensionless common parameter ξ_T can be revealed through sound velocity data obtained from the Einstein–Lindemann model.

Combination of the modified model for ν, represented by Eq. (6.36), and the Einstein model for $U\left(= q_1 \nu V^{1/3}\right)$ gives, at the melting point.

$$U_m = 3.768 \times 10^4 \left(\frac{\gamma_m}{M}\right)^{1/2} V_m^{1/3} \tag{5.22}$$

Eq. (5.22) performs well with Δ(41) and S (41) values of 14.5 per cent and 0.170, respectively,

(f),(g) If we use the relation represented by Eq. (4.10), both the Skapski model and the Schytil model provide exactly the same results of calculations for the surface tension of liquid metallic elements.

(h) Accurate viscosity values may be predicted using the following equation:

$$\mu_m = \frac{3.77 \times 10^2}{k_0} \left(\frac{M}{\xi_T T_m}\right)^{1/2} \gamma_m$$

where k_0 is the dimensionless numerical factor. The values of k_0 for all the metallic elements can be obtained using the relationship in Figure 6.6.

A detailed explanation is described in Chapter 15.

(i) The values of the Andrade coefficient C_A vary periodically with atomic number. Using the periodic relationship, more accurate viscosity values may be predicted.

(j) With $\xi = 5.32$, the modified Stokes–Einstein formula provides accurate self-diffusivities of liquid metallic elements, giving Δ(18) and S(18) values of 7.8 per cent and 0.102, respectively.

(k) At, or near, the melting point temperatures, the Wiedemann–Franz–Lorenz law, to a good approximation, holds for a number of liquid metals. This law is given on the free electron model (for electronic conduction only), and moreover, on the assumption that the relaxation times, or the mean free times, are identical for electrical and thermal processes.

Einstein–Lindemann model, and modified models in terms of $\xi_{\mathrm{T}}^{1/2}$ or $\xi_{\mathrm{E}}^{1/2}$ for several thermophysical properties of liquid metallic elements at their melting point temperatures.

In the 1990s, Egry [16] and Yokoyama [17], independently, have both pointed out, in their research articles on the surface tension, viscosity, and self-diffusion of liquid metals, that there is a need for correction factors in order to accurately predict such property values.

To assess the performance of the Fowler–Born–Green relation (see Subsection 7.2.1), Egry [16] has introduced a parameter, denoted by Q. The parameter Q is defined as

$$Q = \frac{16}{15}\left(\frac{m}{kT}\right)^{1/2}\frac{\gamma}{\mu} \tag{10.3}$$

If the Fowler–Born–Green relation holds strictly, the value of Q is equal to unity (see Table 10.3). Egry [16] and Egry et al. [18] investigated the validity of the Fowler–Born–Green relation, on the basis of Eq. (10.3), for six selected metals, i.e. liquid iron group and copper group metals; the results of calculations are iron (0.82), cobalt (1.02), nickel (0.81), copper (0.99), silver (0.82), and gold (0.98), where the numerical figures in parentheses represent the values of Q at the melting point temperatures.

Yokoyama [17] calculated the viscosities of iron group and copper group metals, using the Stokes–Einstein formula, namely

$$\mu = \frac{kT}{2\pi\sigma D_{\mathrm{HS}}} \tag{10.4}$$

where D_{HS} is the self-diffusivity in a hard-sphere fluid, σ being its hard-sphere diameter. According to Yokoyama, the agreement between theory and experiment for their viscosity values is reasonable. However, Yokoyama has indicated that the quality of the agreement is greatly improved by considering a correction factor, denoted by C, suggested in the following equation

$$\mu = \frac{kT}{C2\pi\sigma D_{\mathrm{HS}}} \tag{10.5}$$

and, further, indicated that the correction factor C is exactly the same as Q introduced by Egry [16]. Combining Eqs. (6.12) and (10.5), Yokoyama [17] obtained:

$$C = \frac{16}{15}\left(\frac{m}{kT}\right)^{1/2}\frac{\gamma}{\mu} = Q$$

Both these approaches suggest that a new correction factor Q, or C, is needed for accurate predictions of a liquid metal's surface tension, or viscosity (or self-diffusivity). To the authors' knowledge, however, neither Egry nor Yokoyama gave further discussions on these correction factors.

Combination of Eqs. (7.41) and (10.3) (i.e. Eq. (10.3) at the melting point temperature) gives

$$Q = C = 9.810k_0\xi_T^{1/2} \tag{10.6}$$[2]

The value of k_0 is about 0.10 for liquid metals, so that the correction factors, Q and C, are roughly the same as $\xi_T^{1/2}$ introduced by Iida et al.

In the subsequent six chapters, predictive models, expressed in terms of $\xi_T^{1/2}$ or $\xi_E^{1/2}$, for several thermophysical properties of liquid metallic elements are discussed from the viewpoint of materials process science. Since, in the field of materials process science, both accuracy and university are required of any model for the thermophysical properties of metallic liquids, we now make mention of the following point. In order to accurately predict values of several thermophysical properties of liquid metallic elements, empirical relations (e.g. the periodic variation of physical quantity values with atomic number) based on experimental data are frequently used in this book. Such an approach depends on the assumptions that all the experimental data used are reliable, and that experimental uncertainties of the data are properly estimated.

10.4 Periodic Table and Grouping, or Classification, of the Chemical Elements

As mentioned in Subsection 1.6.2, the thermophysical properties of simple substances are essentially determined by the configurations of electrons, especially of the outermost electrons. The modern periodic table of the chemical elements, or simply the 'elements', is based on the electron configurations of atoms. Table 10.4 shows one form of periodic table in which the grouping, or the classification, of the chemical elements in this book is given (cf. also Tables 1.6 and 1.7 and Appendix 1).

10.5 Periodic Variations in Values of Some 12 Physical Quantities or Thermophysical Properties with Atomic Number

It has long been known that the physical and chemical properties of elements obey the periodic law.[3] Figures 10.2 to 10.13 show values for various physical quantities or thermophysical properties of elements plotted against atomic number;[4] their data are

[2] Using Eq. (4.10), we have: $Q = C = 0.523k_0\xi_E^{1/2}(\Delta_l^g H_b / T_m)^{1/2}$.

[3] The periodic law states that the physical and chemical properties of elements are a periodic function of their atomic number (or their proton number).

[4] Values for various physical quantities or thermophysical properties of liquid metallic elements are plotted, excluding several figures.

Table 10.4 *Grouping, or classification, of the chemical elements in this book.*

Phase (under standard conditions)

g	gas
l	liquid
m	metal
sm	semimetal
sc	semiconductor

Key: each box shows phase, chemical symbol and atomic number, e.g. g H 1 (Chemical symbol: H; Atomic number: 1).

Alkali metals (Group IA elements, with the exception of hydrogen)

Transition elements* (Groups 4–11)

Early transition elements (Groups IVB–VIIB, 4–7)

Copper group (transition) elements (Group IB, 11)

Zinc group elements (Group IIB, 12)

Period \ Group	IA 1	IIA 2	IIIB 3	IVB 4	VB 5	VIB 6	VIIB 7	VIIIB 8	VIIIB 9	VIIIB 10	IB 11	IIB 12	IIIA 13	IVA 14	VA 15	VIA 16	VIIA 17	VIIIA 18
1	g H 1																	g He 2
2	m Li 3	m Be 4											sc B 5	sc C 6	g N 7	g O 8	g F 9	g Ne 10
3	m Na 11	m Mg 12											m Al 13	sc Si 14	sc P 15	sc S 16	g Cl 17	g Ar 18
4	m K 19	m Ca 20	m Sc 21	m Ti 22	m V 23	m Cr 24	m Mn 25	m Fe 26	m Co 27	m Ni 28	m Cu 29	m Zn 30	m Ga 31	sc Ge 32	sm As 33	sc Se 34	l Br 35	g Kr 36
5	m Rb 37	m Sr 38	m Y 39	m Zr 40	m Nb 41	m Mo 42	m Tc 43	m Ru 44	m Rh 45	m Pd 46	m Ag 47	m Cd 48	m In 49	m Sn 50	sm Sb 51	sc Te 52	sc I 53	g Xe 54
6	m Cs 55	m Ba 56	m La 57 †	m Hf 72	m Ta 73	m W 74	m Re 75	m Os 76	m Ir 77	m Pt 78	m Au 79	l,m Hg 80	m Tl 81	m Pb 82	sm Bi 83	m Po 84	m At 85	g Rn 86
7	m Fr 87	m Ra 88	m Ac 89 ‡															

† Lanthanoid elements (La–Lu)

Rare earth elements	m Sc 21	m Y 39	m La 57	m Ce 58	m Pr 59	m Nd 60	m Pm 61	m Sm 62	m Eu 63	m Gd 64	m Tb 65	m Dy 66	m Ho 67	m Er 68	m Tm 69	m Yb 70	m Lu 71
‡ Actinoid elements	m Ac 89	m Th 90	m Pa 91	m U 92	m Np 93	m Pu 94	m Am 95	m Cm 96	m Bk 97	m Cf 98	m Es 99	m Fm 100	m Md 101	m No 102	m Lr 103		

	Period	VIIIB 8	VIIIB 9	VIIIB 10
Iron group elements	4	m Fe 26	m Co 27	m Ni 28
Platinum group elements	5	m Ru 44	m Rh 45	m Pd 46
Platinum group elements	6	m Os 76	m Ir 77	m Pt 78

* These elements are usually called three 'main transition elements (or series)'.

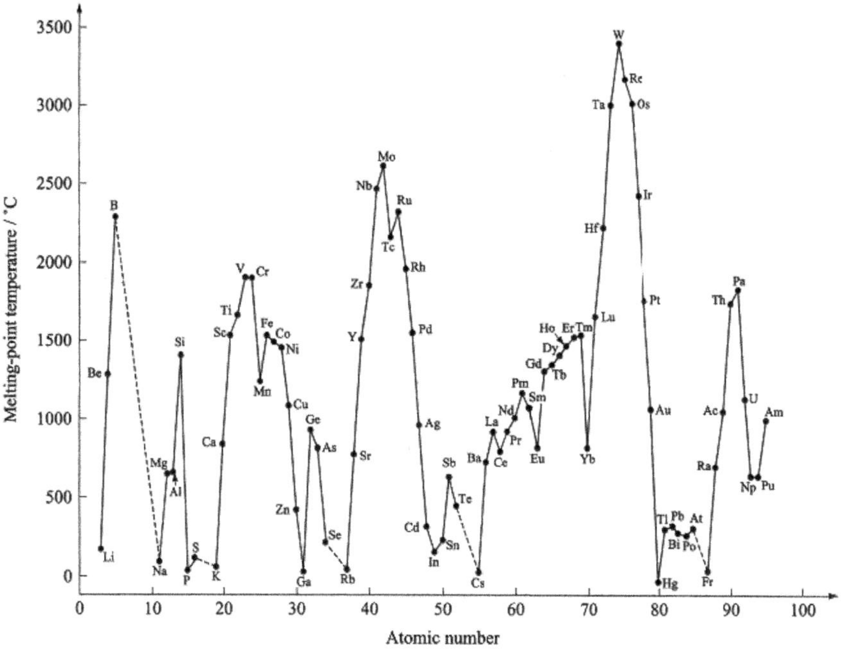

Figure 10.2 *Melting point temperature plotted against atomic number for a large number of metallic elements (i.e. metals, semimetals, and semiconductors; see Table 10.4).*

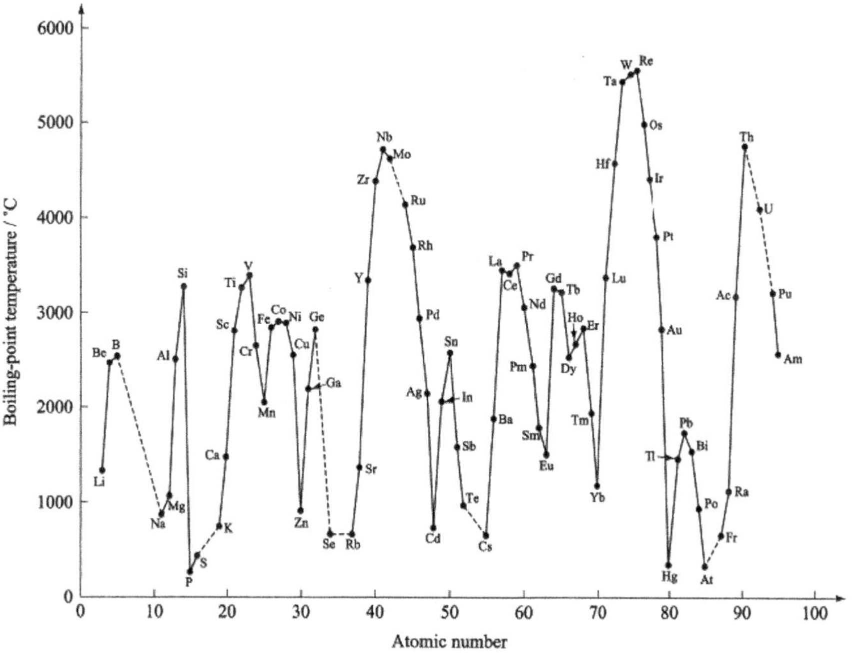

Figure 10.3 *Boiling point temperature plotted against atomic number for a large number of metallic elements.*

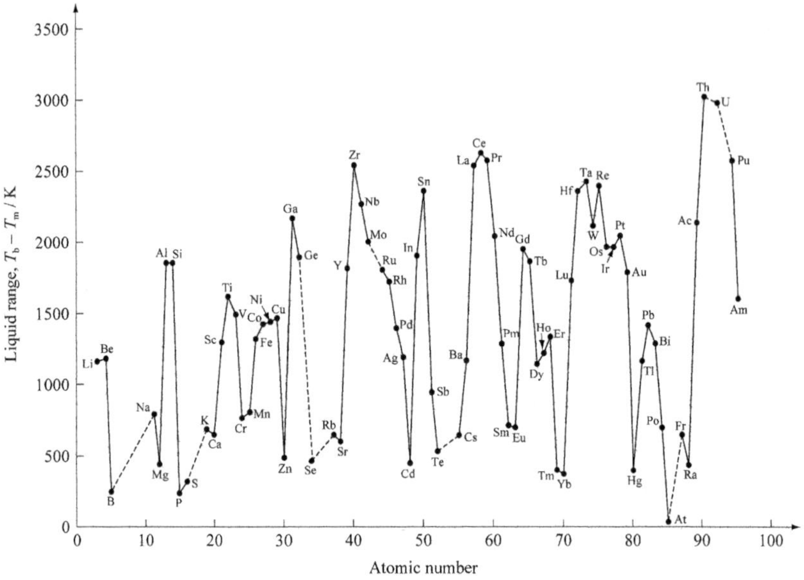

Figure 10.4 *Liquid range (i.e. $T_b - T_m$) plotted against atomic number for a large number of metallic elements.*

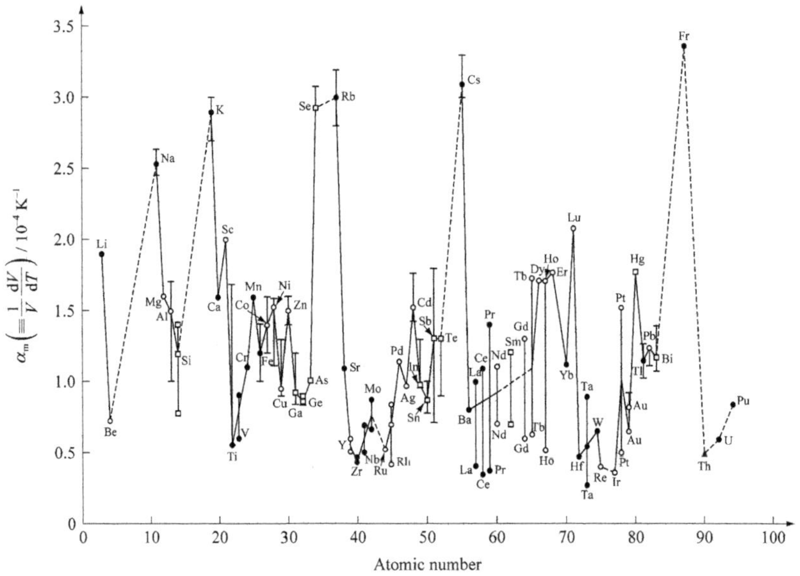

Figure 10.5 *Volume expansivity plotted against atomic number (cf. Figure 3.9) for a large number of liquid metallic elements at their melting points. The symbols refer to the crystal structures of solid metallic elements at their melting points: ○, close-packed structures; ●, ▲, body-centred cubic structure (▲, estimated by Steinberg [8]); □, other complex structures. The bars indicate the extremes of experimental values which are estimated to be reasonable. Points linked by a vertical line represent two or three different experimental values for a single metallic element.*

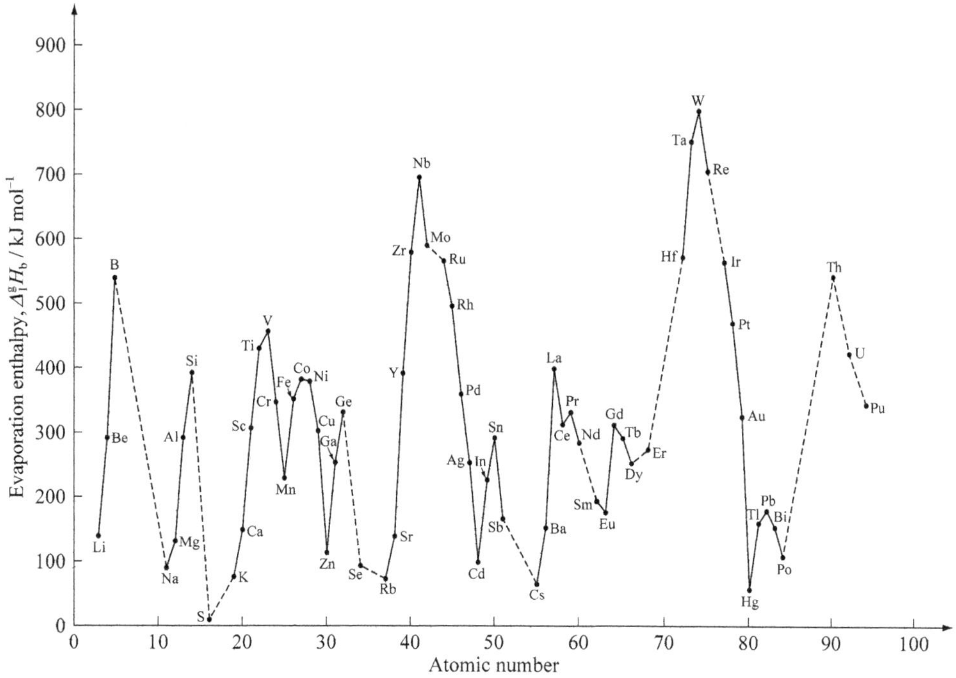

Figure 10.6 *Evaporation enthalpy plotted against atomic number for a large number of liquid metallic elements at their boiling points.*

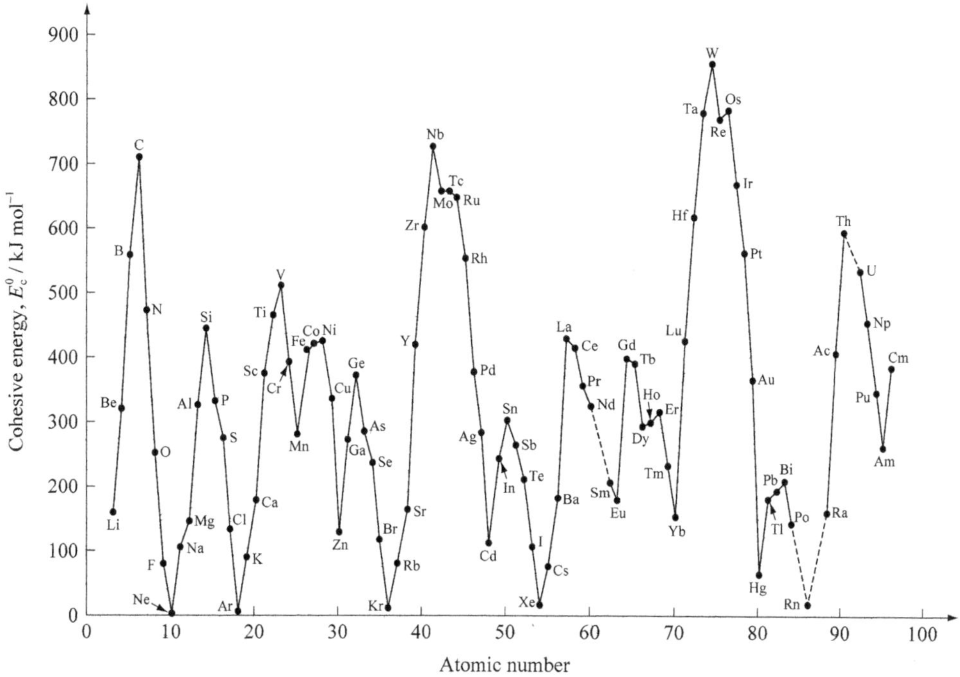

Figure 10.7 *Cohesive energy at 0 K at 1 atm (101.325 kPa) plotted against atomic number for a large number of solid elements (see Table 4.1).*

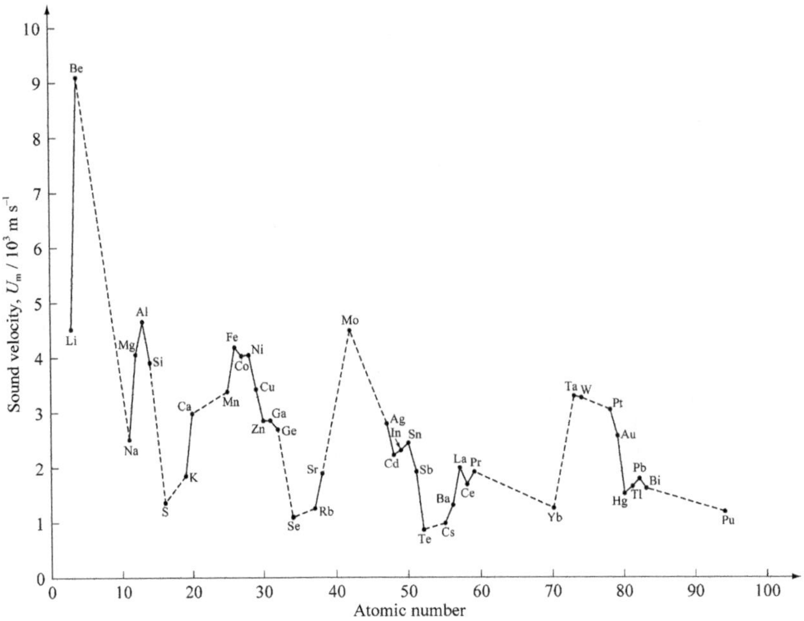

Figure 10.8 *Sound velocity in various liquid metallic elements, plus sulphur, at their melting points against atomic number.*

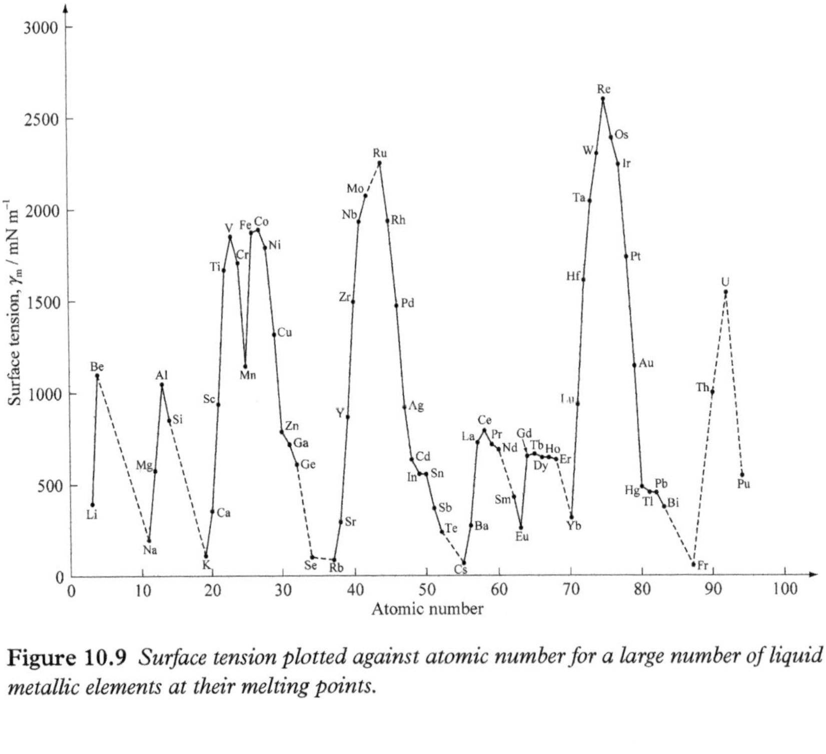

Figure 10.9 *Surface tension plotted against atomic number for a large number of liquid metallic elements at their melting points.*

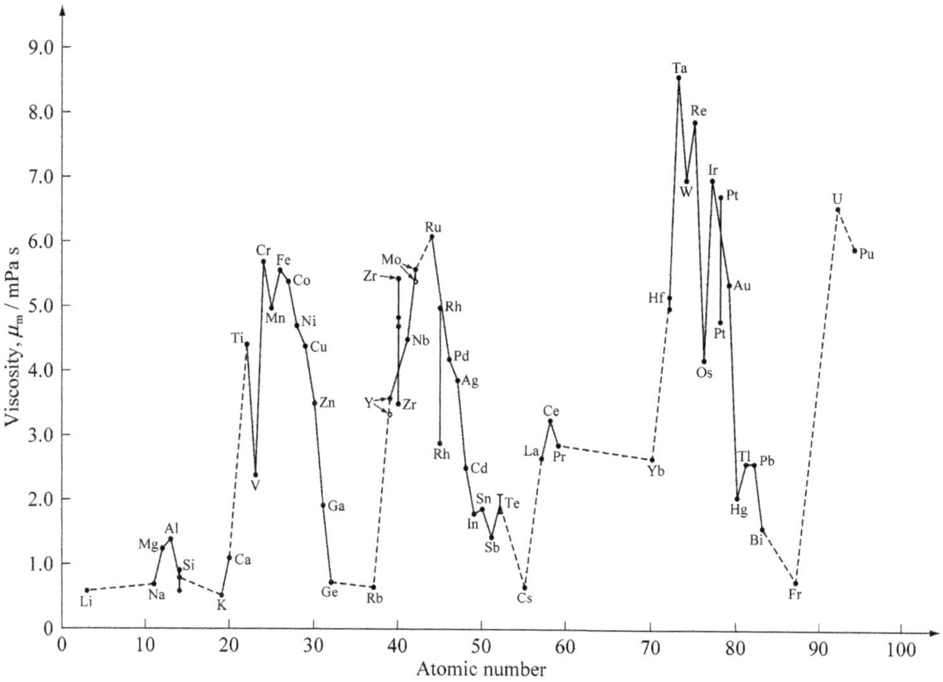

Figure 10.10 *Viscosity of liquid metallic elements at their melting points plotted against atomic number: ●, experimental, and ○, predicted values [19]. Points linked by a vertical line represent two, three, or four different experimental values for a single metallic element.*

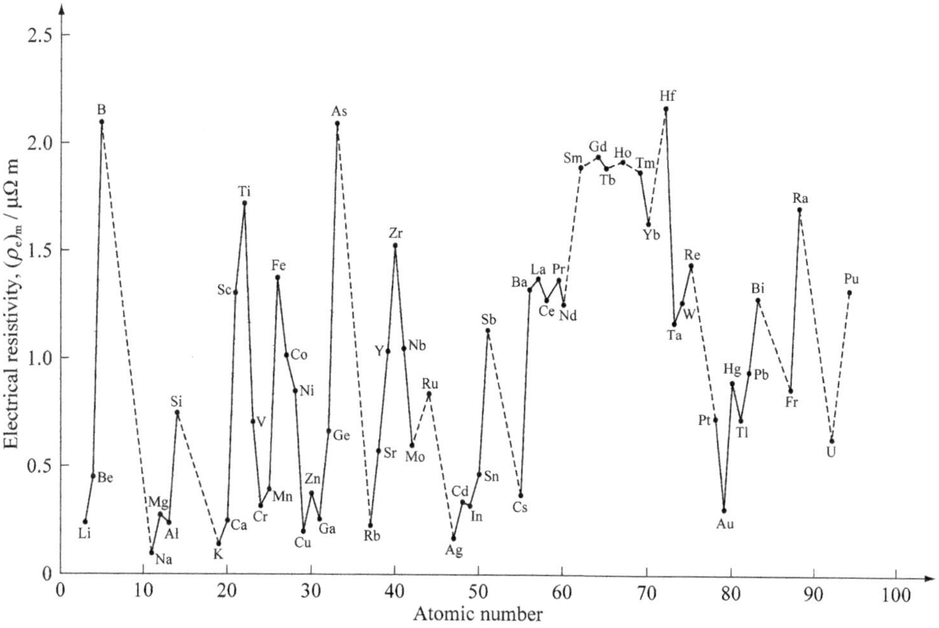

Figure 10.11 *Electrical resistivity plotted against atomic number for a large number of liquid metallic elements at their melting points.*

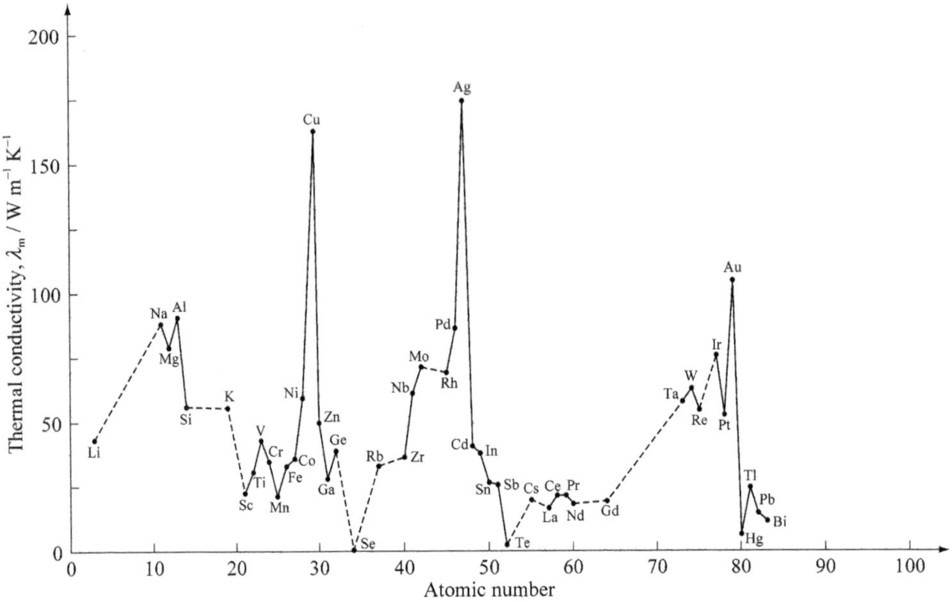

Figure 10.12 *Thermal conductivity of liquid metallic elements at their melting points plotted against atomic number.*

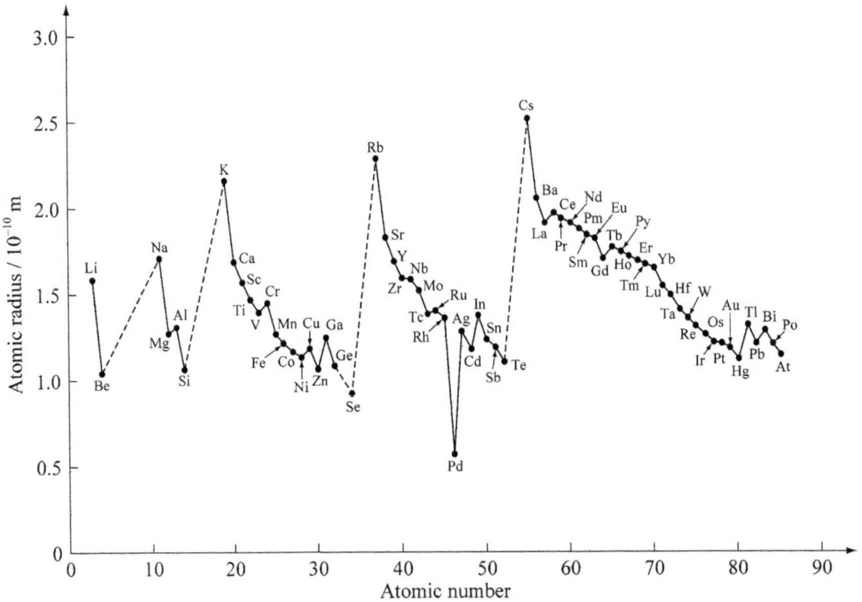

Figure 10.13 *Theoretically calculated radius plotted against atomic number for a separated neutral atom (i.e. the distance from the centre of the nucleus to the outermost electron, or to the maximum electron density in the outermost shell). Data are taken from Mizoguchi [20].*

given in Chapters 1, 4, 9, and 17. The periodic law generally provides rough predictions for the thermophysical property values of elements, and further, is helpful in obtaining information on their anomalous properties or behaviour. For example, as can be seen from Figure 10.2, in the *d*-block transition metals in period 4, Ti(1668) → V(1910) → Cr(1907) → Mn(1246) → Fe(1538) → Co(1495) → Ni(1455) → Cu(1085), where the numerical figures in parentheses denote melting point in °C; evidently, manganese shows an anomalous discontinuity (or a sharp variation) in the melting point temperature plots; in the lanthanoid series, cerium, europium, and ytterbium similarly show anomalous discontinuities (see also Figure 10.9).

REFERENCES

1. F. Schytil, *Z. Naturforsh.*, **4** (1949), 191.
2. B.J. Keene, *Int. Mater. Rev.*, **38** (1993), 157.
3. K.C. Mills and Y.C. Su, *Int. Mater. Rev.*, **51** (2006), 329.
4. T. Ishikawa and P.-F. Paradis, in *High-Temperature Measurements of Materials*, H. Fukuyama and Y. Waseda (eds.), Springer, Berlin, 2009, p.186.
5. E.N. da C. Andrade, *Phil. Mag.*, **17** (1934), 497.
6. F. Trouton, *Nature*, **27** (1883), 292; F. Trouton, *Phil. Mag.*, **18** (1884), 54.
7. F.A. Lindemann, *Phys. Z.*, **11** (1910), 609.
8. D.J. Steinberg, *Met. Trans.*, **5** (1974), 1341.
9. M.B. Gitis and I.G. Mikhailov, *Sov. Phys. Acoust.*, **13** (1968), 473.
10. A. Einstein, *Ann. Phys.*, **34** (1911), 170.
11. A.S. Skapski, *J. Chem. Phys.*, **16** (1948), 386; A.S. Skapski, *J. Chem. Phys.*, **16** (1948), 389.
12. R.H. Fowler, *Proc. R. Soc. Lond., A*, **159** (1937), 229.
13. M. Born and H.S. Green, *Proc. R. Soc. Lond., A*, **190** (1947), 455.
14. H. Eyring and T. Ree, *Proc. Natl. Acad. Sci.*, **47** (1961), 526.
15. L. Lorenz, *Ann. Phys. Chem.*, **147** (1872), 429.
16. I. Egry, *Scr. Metall. Mater.*, **26** (1992), 1349.
17. I. Yokoyama, *Physica B*, **271** (1999), 230.
18. I. Egry, G. Lohöfer, and S. Sauerland, *J. Non-Cryst. Solids*, **156–158** (1993), 830.
19. T. Iida, R. Guthrie, M. Isac, and N. Tripathi, *Metall. Mater., Trans. B*, **37** (2006), 403.
20. T. Mizoguchi, *Fundamentals of Materials Science: Solid State Physics*, Shokabo, Tokyo, 1989, p. 386.

11

Predictions of the Velocity of Sound

11.1 Introduction

As we have already mentioned in Chapter 5, sound velocities in liquid metallic elements are one of their most basic thermodynamic properties. In addition, it has been gradually revealed that the velocity of sound also gives valuable information on transport, or dynamic, properties. Useful common parameters, or dimensionless numbers, for better predictions of several thermophysical properties can easily be extracted from the sound velocities in liquid metallic elements. Unfortunately, experimentally derived data for sound velocities in liquid metallic elements are relatively scanty; at present, it seems such data are available only for approximately 40 liquid metallic elements. In fact, most transition and rare earth metals have never been measured experimentally. Thus, the authors have developed two models for accurate predictions of the sound velocity in liquid metallic elements at their melting point temperatures. These models are expressed in terms of a few well-known physical quantities.

In this chapter, the authors predict values of the melting point sound velocity in various liquid metals for which experimental data will probably still be lacking. Armed with these predicted sound velocities, the common parameters are determined for these liquid metals.

11.2 Predictions of Melting Point Sound Velocities in Various Liquid Metals

The 41 data points plotted in Figure 5.2 (i.e. the modified Einstein–Lindemann model for the sound velocities in liquid metallic elements, represented by $U_m = 3.768 \times 10^4 (\gamma_m / M)^{1/2} V_m^{1/3})$ can roughly be divided into three groups. This is shown in Figure 11.1; the three groups and their outlines are given in Table 11.1. In a word, the data points of the groups 1 and 2 are, respectively, separable from the total data points. Using $3.768 \times 10^4 \text{ mol}^{-1/6}$ as the base slope (see Figure 11.1), differences in the slope of $3.083 \times 10^4 \text{ mol}^{-1/6}$ (group 1) and $4.440 \times 10^4 \text{ mol}^{-1/6}$ (group 2) are −18.2 and 17.8 per cent, respectively. These differences are much larger than the uncertainties in the experimental values (typically ± 2 to ± 3 per cent) for the melting

The Thermophysical Properties of Metallic Liquids: Volume 2 – Predictive Models. First Edition.
Takamichi Iida and Roderick I. L. Guthrie. © Takamichi Iida and Roderick I. L. Guthrie 2015.
Published in 2015 by Oxford University Press.

point sound velocity in liquid metallic elements. This grouping may be mainly due to a rough approximation of the correction factor β for Lindemann's melting equation (see Subsections 1.2.2.4 and 6.3.2).

Combination of Eqs. (6.37) and (7.14) gives

$$U_{\mathrm{m}} = \frac{\gamma_{\mathrm{m}} V_{\mathrm{m}}^{2/3}}{k_0 C_{\mathrm{A}} (M T_{\mathrm{m}})^{1/2}} = \frac{1}{\gamma_0}\left(\frac{\gamma_{\mathrm{m}}}{M}\right)^{1/2} V_{\mathrm{m}}^{1/3} \cdot \frac{\gamma_{\mathrm{m}}^{1/2} V_{\mathrm{m}}^{1/3}}{T_{\mathrm{m}}^{1/2}}$$

$$\gamma_0 \equiv k_0 C_{\mathrm{A}}$$

From Eq. (6.51) (i.e. $\gamma_{\mathrm{m}} \simeq 6.8 \times 10^{-8} R T_{\mathrm{m}} / V_{\mathrm{m}}^{2/3}$) we have

$$\gamma_{\mathrm{m}}^{1/2} \simeq 7.5 \times 10^{-4} \frac{T_{\mathrm{m}}^{1/2}}{V_{\mathrm{m}}^{1/3}}$$

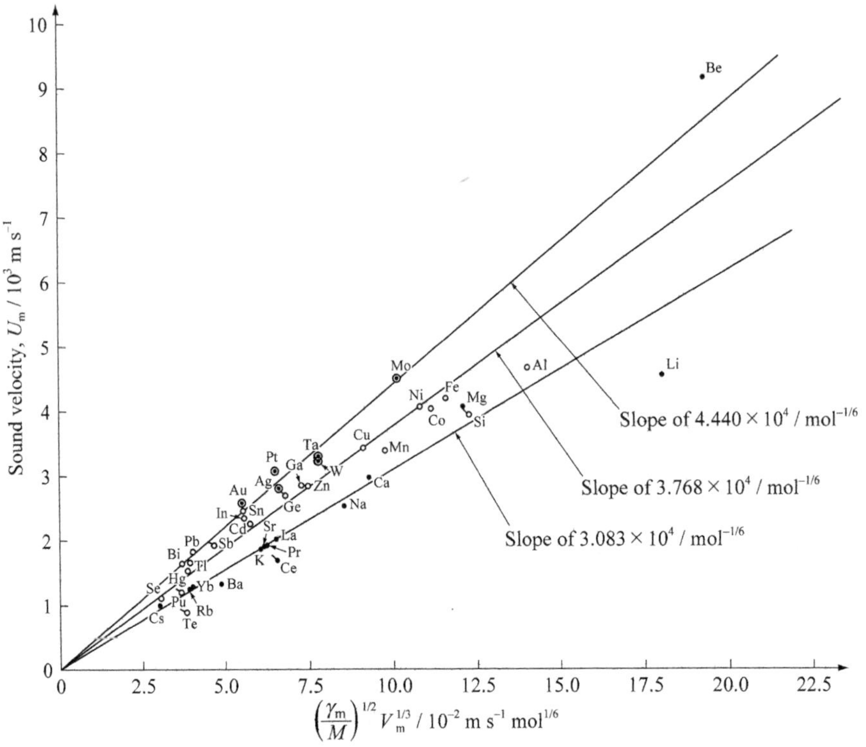

Figure 11.1 *Melting point sound velocity in various liquid metallic elements as a function of $(\gamma_m / M)^{1/2} V_m^{1/3}$, i.e. the modified Einstein–Lindemann model.* ●*, s-block and rare earth metals (group 1);* ⊙*, d-block transition metals in periods 5 and 6 (group 2);* ○*, other metals (excluding groups 1 and 2) (group 3)* ○, ●*, show positive temperature coefficients (see Table 17.7(b)).*

Table 11.1 *Grouping of metals, and equations, for predicting melting point sound velocities in liquid metals.*

Group	Metal which belongs to the group	Metal to be predicted	Equation
1	IA, IIA, and rare earth metals (excluding Li and Be)	Fr, Dy, Er, Eu, Gd, Ho, Nd, Sm, Sc, Tb, Y (excluding Lu)	$3.083 \times 10^4 \left(\frac{\gamma_m}{M}\right)^{1/2} V_m^{1/3}$
2	Transition metals in periods 5 through 7	Hf, Ir, Nb, Os, Pd, Re, Rh, Ru, Th, U, Zr	$4.440 \times 10^4 \left(\frac{\gamma_m}{M}\right)^{1/2} V_m^{1/3}$
3	Other metals (excluding the Groups 1 and 2)	Cr, Ti, V, Lu	$\frac{1}{2}\left\{3.768 \times 10^4 \left(\frac{\gamma_m}{M}\right)^{1/2} V_m^{1/3} + \frac{5.579}{M^{1/4} V_m^{1/2}}\right\}^{\dagger}$

† The mean value of sound velocities predicted from Eqs. (5.22) and (5.40).

or

$$\frac{\gamma_m^{1/2} V_m^{1/3}}{T_m^{1/2}} \simeq 7.5 \times 10^{-4}$$

Thus

$$U_m \simeq \frac{7.5 \times 10^{-4}}{\gamma_0} \left(\frac{\gamma_m}{M}\right)^{1/2} V_m^{1/3} \simeq 3.9 \times 10^4 \left(\frac{\gamma_m}{M}\right)^{1/2} V_m^{1/3}$$

$$(\gamma_0 \simeq 1.9 \times 10^{-8}, \text{ see Figure 14.2})$$

This equation is very similar to the modified Einstein–Lindemann model proposed in Chapter 5.

In order to accurately calculate or predict values for metallic liquid viscosities, surface tensions, and sound velocities, the parameter γ_0, or $k_0 C_A$, is of crucial importance (cf. Figures 14.2, 15.1, and 15.6).

Equations used for predicting the melting point sound velocities U_m in liquid metals for the three groups of metals are expressed in terms of three well-known physical quantities, as follows:

$$\text{For group 1,} \quad U_m = 3.083 \times 10^4 \left(\frac{\gamma_m}{M}\right)^{1/2} V_m^{1/3} \tag{11.1}$$

$$\text{For group 2,} \quad U_m = 4.440 \times 10^4 \left(\frac{\gamma_m}{M}\right)^{1/2} V_m^{1/3} \tag{11.2}$$

where the numerical factors of 3.083×10^4 mol$^{-1/6}$ and 4.440×10^4 mol$^{-1/6}$ were determined so as to give the minimum values of relative standard deviation S for the liquid metallic elements in groups 1 and 2, respectively. This minimization approach is also used in the evaluation of the numerical factors of Eqs. (11.4) through (11.6)

$$\text{For group 3,}\quad U_{\text{m}} = \frac{1}{2}\left\{3.768 \times 10^4 \left(\frac{\gamma_{\text{m}}}{M}\right)^{1/2} V_{\text{m}}^{1/3} + \frac{5.579}{M^{1/4} V_{\text{m}}^{1/2}}\right\} \tag{11.3}$$

On the basis of the results in Table 5.7, the mean values of the sound velocities calculated from Eqs. (5.22) and (5.40) are adopted for group 3.

Table 11.2 shows the performances of the equations which are to be used for predicting the melting point sound velocities in liquid metals for groups 1 through 3, respectively; following the result of the group 3, the same approach (as the group 3) to the groups 1 and 2 are also given for reference. Incidentally, lithium and beryllium are excluded from group 1. As seen from Figure 11.1, they (i.e. lithium and beryllium) show large deviations from the relationship represented by Eq. (11.1). Further, the maximum ordinate of the metals which are to be predicted in the group 1 is 11.70 / 10^{-2} m s^{-1} mol$^{1/6}$ for scandium, so that the data points of both lithium and beryllium are not necessarily needed. Similarly, six metallic elements are excluded from group 3; they, except for selenium,[1] show positive temperature coefficients below certain temperatures, then pass through maxima, and subsequently display negative coefficients with increasing temperature. The metals to be predicted in the group 3 are expected not to exhibit such anomalous behaviour in the velocity of sound. As is obvious from Table 11.2, all of the equations for the three groups perform excellently, giving Δ and S values of 4.8, 4.3, and 5.6 per cent, and 0.069, 0.048, and 0.072, respectively. In addition, accurate, or relatively accurate, values of the physical quantities appearing in Eqs. (11.1) through (11.3) i.e. γ_{m}, M, and V_{m} (or ρ_{m}), are available for a large number of metallic elements (cf. Chapter 17). From this point of view, the Einstein–Lindemann model is also expressed in terms of well-known physical quantities that have accurate values, i.e. T_{m} and M. Equations for the three groups based on the Einstein–Lindemann model are, respectively, given by

$$\text{For group 1,}\quad U_{\text{m}} = 6.750\left(\frac{RT_{\text{m}}}{M}\right)^{1/2} \tag{11.4}$$

$$\text{For group 2,}\quad U_{\text{m}} = 9.388\left(\frac{RT_{\text{m}}}{M}\right)^{1/2} \tag{11.5}$$

and

$$\text{For group 3,}\quad U_{\text{m}} = 9.197\left(\frac{RT_{\text{m}}}{M}\right)^{1/2} \tag{11.6}$$

[1] Values for $(\partial U / \partial T)_p$ of liquid selenium and sulphur cease to be constant with increasing temperature.

Table 11.2 *Performances (i.e. δ_i, Δ, and S values) of Eqs. (11.1) through (11.3), with groups 1 through 3.*

Group 1

Metal		$(U_m)_{exp.}$ m s^{-1}	$(U_m)_{cal.}$ m s^{-1} Eq. (11.1)	δ_i %
Barium	Ba	1331	1503	−11.4
Caesium	Cs	983	925	6.3
Calcium	Ca	2978	2862	4.1
Cerium	Ce	1693	2023	−16.3
Lanthanum	La	2022	2016	0.3
Magnesium	Mg	4065	3728	9.0
Potassium	K	1876	1870	0.3
Praseodymium	Pr	1925	1927	−0.1
Rubidium	Rb	1251	1220	2.5
Sodium	Na	2526	2632	−4.0
Strontium	Sr	1902	1888	0.7
Ytterbium	Yb	1274	1239	2.8
			Δ (12) %	4.8
			S (12)	0.069

Group 2

Metal		$(U_m)_{exp.}$ m s^{-1}	$(U_m)_{cal.}$ m s^{-1} Eq. (11.2)	δ_i %
Gold	Au	2568	2410	6.6
Molybdenum	Mo	4502	4498	0.1
Platinum	Pt	3053	2892	5.6
Silver	Ag	2790	2945	−5.3
Tantalum	Ta	3303	3431	−3.7
Tungsten	W	3279	3436	−4.6
			Δ (6) %	4.3
			S (6)	0.048

continued

Table 11.2 *(continued)*

Group 3

Metal		$(U_m)_{exp.}$ m s^{-1}	$(U_m)_{cal}$ / m s^{-1}			δ_i / %		
			Eq. (5.22)	Eq. (5.40)	Eq. (11.3)†	Eq. (5.22)	Eq. (5.40)	Eq. (11.3)†
Aluminium	Al	4680	5277	4093	4685	–11.3	14.3	–0.1
Bismuth	Bi	1640	1401	1810	1606	17.1	–9.4	2.1
Cadmium	Cd	2237	2162	2575	2369	3.5	–13.1	–5.6
Cobalt	Co	4031	4204	4110	4157	–4.1	–1.9	–3.0
Copper	Cu	3440	3426	3943	3685	0.4	–12.8	–6.6
Gallium	Ga	2873	2733	3216	2975	5.1	–10.7	–3.4
Indium	In	2320	2109	2374	2242	10.0	–2.3	3.5
Iron	Fe	4200	4361	4073	4217	–3.7	3.1	–0.4
Lead	Pb	1821	1504	1877	1691	21.1	–3.0	7.7
Manganese	Mn	3381	3659	3731	3695	–7.6	–9.4	–8.5
Mercury	Hg	1511	1440	2178	1809	4.9	–30.6	–16.5
Nickel	Ni	4047	4066	4160	4113	–0.5	–2.7	–1.6
Thallium	Tl	1650	1480	1955	1718	11.5	–15.6	–4.0
Tin	Sn	2464	2095	2305	2200	17.6	6.9	12.0
Zinc	Zn	2850	2814	3498	3156	1.3	–18.5	–9.7
					Δ (15) %	8.0	10.3	5.6
					S (15)	0.102	0.127	0.072

† The mean value of sound velocities calculated from Eqs. (5.22) and (5.40).

Group 1

Metal		$(U_m)_{exp.}$ m s^{-1}	$(U_m)_{cal}$ / m s^{-1}			δ_i / %		
			Eq. (5.22)	Eq. (5.40)	Eq. (11.3)†	Eq. (5.22)	Eq. (5.40)	Eq. (11.3)†
Barium	Ba	1331	1837	1425	1631	–27.5	–6.6	–18.4
Caesium	Cs	983	1130	1087	1109	–13.0	–9.6	–11.4
Calcium	Ca	2978	3498	2301	2900	–14.9	29.4	2.7
Cerium	Ce	1693	2473	1992	2233	–31.5	15.0	–24.2
Lanthanum	La	2022	2464	1892	2178	–17.9	6.9	–7.2
Magnesium	Mg	4065	4556	3614	4085	–10.8	12.5	–0.5
Potassium	K	1876	2285	1824	2055	–17.9	2.9	–8.7
Praseodymium	Pr	1925	2355	1973	2164	–18.3	–2.4	–11.0
Rubidium	Rb	1251	1491	1358	1425	–16.1	–7.9	–12.2
Sodium	Na	2526	3216	2877	3047	–21.5	–12.2	–17.1
Strontium	Sr	1902	2308	1685	1997	–17.6	12.9	–4.8
Ytterbium	Yb	1274	1514	1703	1609	–15.9	–25.2	–20.8
					Δ (12) %	18.6	12.0	11.6
					S (12)	0.194	0.143	0.136

† The mean value of sound velocities calculated from Eqs. (5.22) and (5.40).

Table 11.2 *(continued)*

Group 2

Metal		$(U_m)_{exp.}$ m s^{-1}	$(U_m)_{cal}$ / m s^{-1}			δ_i / %		
			Eq. (5.22)	Eq. (5.40)	Eq. (11.3)†	Eq. (5.22)	Eq. (5.40)	Eq. (11.3)†
Gold	Au	2568	2046	2485	2266	25.5	3.3	13.3
Molybdenum	Mo	4502	3817	3124	3471	17.9	44.1	29.7
Platinum	Pt	3053	2454	2613	2534	24.4	16.8	20.5
Silver	Ag	2790	2499	2858	2679	11.6	−2.4	4.1
Tantalum	Ta	3303	2912	2459	2686	13.4	34.3	23.0
Tungsten	W	3279	2916	2642	2779	12.4	24.1	18.0
					Δ (6) %	17.5	20.8	18.1
					S (6)	0.184	0.258	0.198

† The mean value of sound velocities calculated from Eqs. (5.22) and (5.40).

Table 11.3 *Performances (i.e. δ_i, Δ, and S values) of the Einstein–Lindmann model, represented by Eqs. (11.4) through (11.6), with groups 1 through 3.*

Group 1

Metal		$(U_m)_{exp.}$ m s^{-1}	$(U_m)_{cal.}$ / m s^{-1} Eq. (11.4)	δ_i %
Barium	Ba	1331	1661	−19.9
Caesium	Cs	983	928	5.9
Calcium	Ca	2978	3247	−8.3
Cerium	Ce	1693	1702	−0.5
Lanthanum	La	2022	1802	12.2
Magnesium	Mg	4065	3793	7.2
Potassium	K	1876	1806	3.9
Praseodymium	Pr	1925	1799	7.0
Rubidium	Rb	1251	1177	6.3
Sodium	Na	2526	2472	2.2
Strontium	Sr	1902	2131	−10.7
Ytterbium	Yb	1274	1547	−17.6
			Δ (12) %	8.5
			S (12)	0.101

continued

Table 11.3 *(continued)*

Group 2

Metal		$(U_m)_{exp.}$ m s^{-1}	$(U_m)_{cal.}$ / m s^{-1} Eq. (11.5)	δ_i %
Gold	Au	2568	2231	15.1
Molybdenum	Mo	4502	4703	–4.3
Platinum	Pt	3053	2769	10.3
Silver	Ag	2790	2896	–3.7
Tantalum	Ta	3303	3650	–9.5
Tungsten	W	3279	3838	–14.6
			Δ (6) %	9.6
			S (6)	0.106

Group 3

Metal		$(U_m)_{exp.}$ m s^{-1}	$(U_m)_{cal.}$ / m s^{-1} Eq. (11.6)	δ_i %
Aluminium	Al	4680	4932	–5.1
Bismuth	Bi	1640	1354	21.1
Cadmium	Cd	2237	1928	16.0
Cobalt	Co	4031	4593	–12.2
Copper	Cu	3440	3877	–11.3
Gallium	Ga	2873	1748	64.4
Indium	In	2320	1622	43.0
Iron	Fe	4200	4775	–12.0
Lead	Pb	1821	1427	27.6
Manganese	Mn	3381	4409	–23.3
Mercury	Hg	1511	906	66.8
Nickel	Ni	4047	4551	–11.1
Thallium	Tl	1650	1409	17.1
Tin	Sn	2464	1730	42.4
Zinc	Zn	2850	2730	4.4
			Δ (15) %	25.2
			S (15)	0.318

The performances of Eqs. (11.4) through (11.6) are shown in Table 11.3. As is evident from a comparison between Tables 11.2 and 11.3, the modified Einstein–Lindemann model (represented by Eqs. (11.1) through (11.3)) provides much better agreement with experimental values. Generally speaking, models should be as simple as possible, but overly simple models give no accurate predictions (i.e. everything has its limit). Thus, it can safely be said that Eqs. (11.1) through (11.3) are endowed with the necessary conditions of being predictive.

Using Eqs. (11.1) through (11.3), the authors now predict values for the sound velocity in various liquid metals. Metals to be predicted and equations used for predictions of the sound velocities are also summarized in Table 11.1. Table 11.4 lists predicted values for the melting point sound velocity in twenty-six liquid metals. The values of the physical quantities used for calculating their sound velocities are given in Table 11.5. In Figure 11.2, both experimental and predicted values for the melting point sound velocity in liquid metals (i.e. in 77 liquid metallic elements, plus liquid sulphur) have been plotted against atomic number. We see that the sound velocities in liquid metallic elements also obey the periodic law. Higher values are apparent for transition metals in which electrons fill their inner shells after outer shells. The periodic Group VIA elements and IIB metals (i.e. zinc group metals) occupy the valleys and the hollows, respectively, of the plot. Taken all together, the sound velocity values decrease with increasing atomic number, or molar mass, because of atomic (or ionic) motions with propagation of the liquids' compressional waves. For example, in the case of alkali metals:

$$\begin{array}{l} 3 \text{ ———— Atomic number} \\ \text{Li (4554) ——Sound Velocity} / \text{ms}^{-1} \\ 6.941 \text{——Molar mass} / 10^{-3}\ \text{kg mol}^{-1} \end{array}$$

$$\underset{6.941}{\overset{3}{\text{Li}}}\,(4554) \rightarrow \underset{22.990}{\overset{11}{\text{Na}}}\,(2526) \rightarrow \underset{39.098}{\overset{19}{\text{K}}}\,(1876) \rightarrow \underset{85.468}{\overset{37}{\text{Rb}}}\,(1251) \rightarrow \underset{132.905}{\overset{55}{\text{Cs}}}\,(983) \rightarrow \underset{223.020}{\overset{87}{\text{Fr}}}\,(741^{*})$$

where the numerical figure with asterisk * represents a predicted sound velocity value (see Table 11.4).

In 2007, on the basis of Eqs. (5.20), (6.34), and (6.35), Blairs [1,2] examined the velocity of sound U_m in the liquid metallic elements at their melting point temperatures; for a forced line fit through the origin, Blairs obtained the following equation:

$$U_\text{m} = 17.793\beta\left(\frac{RT_\text{m}}{M}\right)^{1/2} \tag{11.7}$$

$$\beta = 1.1 \times 10^3\left\{\frac{V_\text{m}^{1/3}}{(1.97\eta_\text{m}^{1/3} - 1)}\right\}\left(\frac{\gamma_\text{m}}{RT_\text{m}}\right)^{1/2}$$

where η_m is the packing fraction at the melting point; $\eta_\text{m} = 0.463$ is a good approximation for all liquid metals.

From an independent standpoint, Blairs [1–3] proposed the following equation for the melting point sound velocity in liquid metallic elements:

Table 11.4 *Predicted (or calculated) values for the velocity of sound in liquid metals at their melting point temperatures, and values of the dimensionless common parameters,* $\xi_T^{1/2}$ *and* $\xi_E^{1/2}$, *obtained from the predicted sound velocities.*

Metal		(U_m) pre $m\,s^{-1}$	$\xi_T^{1/2}$	$\xi_E^{1/2}$
Chromium	Cr	4216	0.776	1.152
Dysprosium	Dy	1654	0.613	0.941
Erbium	Er	1603	0.582	0.890
Europium	Eu	1257	0.558	0.826
Francium	Fr	741	0.762	–
Gadolinium	Gd	1725	0.648	0.866
Hafnium	Hf	3371	1.073	1.333
Holmium	Ho	1639	0.601	–
Iridium	Ir	3230	1.024	1.333
Lutetium	Lu	2162	0.775	–
Neodymium	Nd	1873	0.746	0.944
Niobium	Nb	4629	1.015	1.196
Osmium	Os	3335	0.954	–
Palladium	Pd	3586	1.032	1.377
Rhenium	Re	3569	0.987	1.295
Rhodium	Rh	4087	1.045	1.316
Ruthenium	Ru	4411	1.036	1.316
Samarium	Sm	1422	0.567	0.890
Scandium	Sc	3608	0.677	0.980
Terbium	Tb	1709	0.637	0.890
Thorium*	Th	2594	1.048	1.198
Titanium	Ti	4270	0.800	1.008
Uranium*	U	2718	1.333	1.442
Vanadium	V	4316	0.786	1.018
Yttrium	Y	2677	0.710	0.900
Zirconium	Zr	4537	1.120	1.270

* Radioactive.

Table 11.5 *Values of the thermophysical properties or physical quantities*† *used for calculating sound velocities and dimensionless parameters.*

Metal		γ_m $N\,m^{-1}$	M $kg\,mol^{-1}$	V_m $10^{-6}\,m^3\,mol^{-1}$	T_m K	$\Delta_l^g H_b$ $10^3\,J\,mol^{-1}$
Chromium	Cr	1.710	0.05200	8.27	2180	348
Dysprosium	Dy	0.648	0.16250	19.4	1685	251
Erbium	Er	0.637	0.16726	18.9	1802	271
Europium	Eu	0.264	0.15197	29.6	1095	176
Francium	Fr	0.062	0.22302	94.9	300	–
Gadolinium	Gd	0.664	0.15725	20.2	1586	312
Hafnium	Hf	1.614	0.17849	16.1	2506	571
Holmium	Ho	0.650	0.16493	19.2	1747	–
Iridium	Ir	2.250	0.19222	9.61	2719	564
Lutetium	Lu	0.940	0.17497	17.9	1936	–
Neodymium	Nd	0.687	0.14424	21.57	1294	284
Niobium	Nb	1.937	0.09291	11.9	2750	696
Osmium	Os	2.400	0.19023	9.46	3306	–
Palladium	Pd	1.482	0.10642	10.14	1828	361
Rhenium	Re	2.610	0.18621	9.90	3459	707
Rhodium	Rh	1.940	0.10291	9.53	2237	496
Ruthenium	Ru	2.260	0.10107	9.27	2607	568
Samarium	Sm	0.430	0.15036	20.3	1347	192
Scandium	Sc	0.939	0.04496	16.8	1814	305
Terbium	Tb	0.669	0.15893	19.7	1629	293
Thorium	Th	1.006	0.23204	22.1	2023	544
Titanium	Ti	1.670	0.04788	11.6	1941	430
Uranium	U	1.552	0.23803	13.78	1408	423
Vanadium	V	1.860	0.05094	9.50	2183	458
Yttrium	Y	0.872	0.08891	21.3	1795	393
Zirconium	Zr	1.500	0.09122	16	2128	582

† Data are given in Chapter 17.

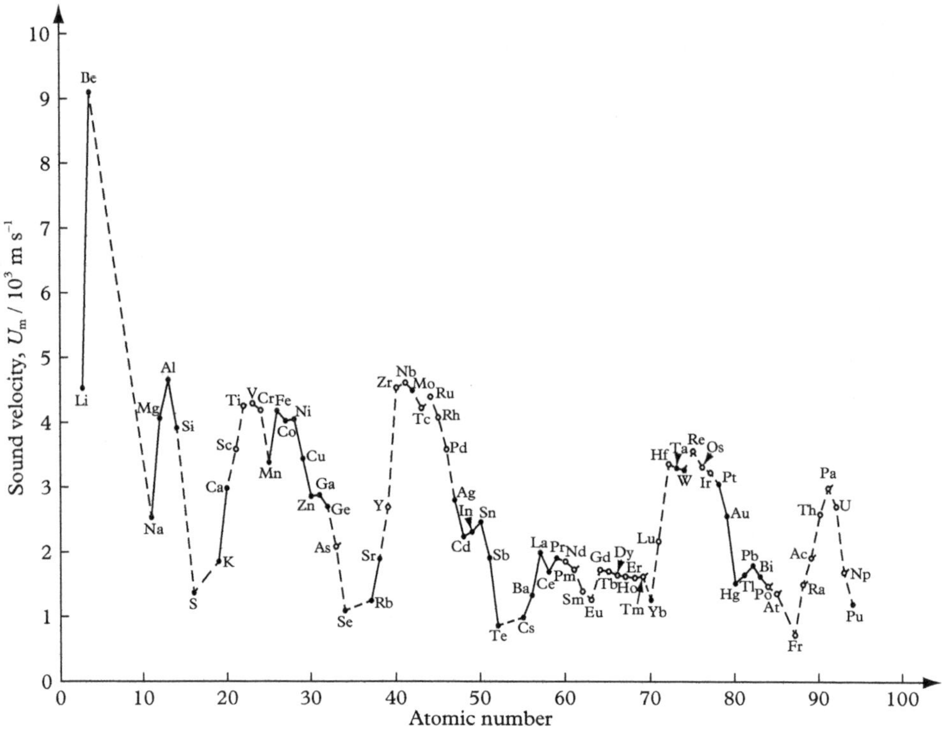

Figure 11.2 *Melting point sound velocity in 77 liquid metallic elements, plus sulphur, plotted against atomic number.* ●, *experimental values;* ○, ♂ *the authors' predicted values (see Tables 11.4 and 14.5).*

$$\log U_{\mathrm{m}} = 0.5226 \log\left(\frac{\gamma_{\mathrm{m}}}{\rho_{\mathrm{m}}}\right) + 5.4364, \text{ (in SI units)} \qquad (11.8)$$

where γ is the surface tension, ρ is the density, and subscript m denotes the melting point.

Table 11.6 lists Blairs' calculated values for the melting point sound velocity. There, for comparison, Blairs included the sound velocities of the rare earth metals calculated by Yokoyama et al. [4] using the Percus–Yevick phonon model of the liquid state, together with low angle diffraction data. Figure 11.3 shows a plot of 68 sound velocity values U_{m} against atomic number, including experimentally derived values, values predicted from Eqs. (11.1), (11.2), (11.3), (11.7), and (11.8), and those calculated by Yokoyama et al. Figure 11.4 shows a plot of the experimental and predicted values (based on Eqs. (11.1) and (11.3)) for lanthanoid metals at their melting point temperatures, against atomic number. As is clear from Figure 11.4, cerium, which shows anomalous temperature dependence of the liquid sound velocity, and the divalent lanthanoids, europium and ytterbium, occupy the valleys of the curves; Blairs' calculated values [2] for liquid lanthanoid metals also display a similar variation with atomic number. Several thermophysical

Table 11.6 *Calculated sound velocities U_m at the melting temperature T_m for liquid transition and rare earth elements (after Blairs [2]).*

Element		Sound velocity U_m / m s^{-1}		
		Eq. (11.7)	Eq. (11.8)	Yokoyama et al. [4]
Chromium	Cr	4116	3663	–
Dysprosium	Dy	1944	1941	2130
Erbium	Er	1882	1867	2450
Europium	Eu	1477	1568	1860
Gadolinium	Gd	2027	2041	2240
Hafnium	Hf	2631	2559	–
Holmium	Ho	1926	1919	2560
Iridium	Ir	2643	2416	–
Lutetium	Lu	2198	2176	2380
Neodymium	Nd	2176	2212	2200
Niobium	Nb	3593	3385	–
Osmium	Os	2777	2487	–
Palladium	Pd	2926	2657	–
Rhenium	Re	2943	2665	–
Rhodium	Rh	3283	2950	–
Ruthenium	Ru	3563	3214	–
Samarium	Sm	1672	1670	–
Scandium	Sc	4237	4272	–
Terbium	Tb	2009	2014	2120
Titanium	Ti	4546	4309	–
Vanadium	V	4683	4255	–
Yttrium	Y	3143	3258	–
Zirconium	Zr	3666	3648	–

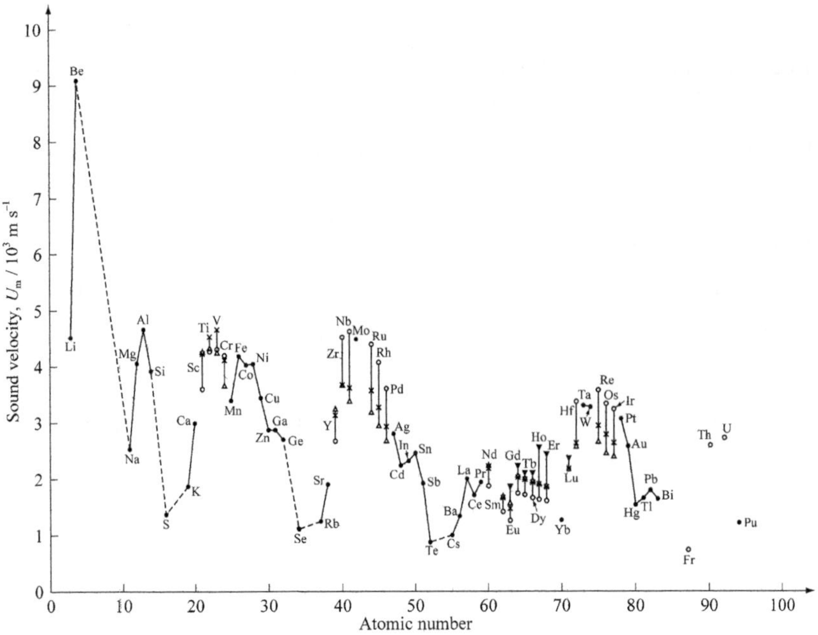

Figure 11.3 *Melting point sound velocity in 67 liquid metallic elements, plus sulphur, plotted against atomic number. ●, experimental values; ○, the authors' predicted values; ×, △, Blairs' predicted values; ▼, Yokoyama et al.'s predicted values (see Table 11.6).*

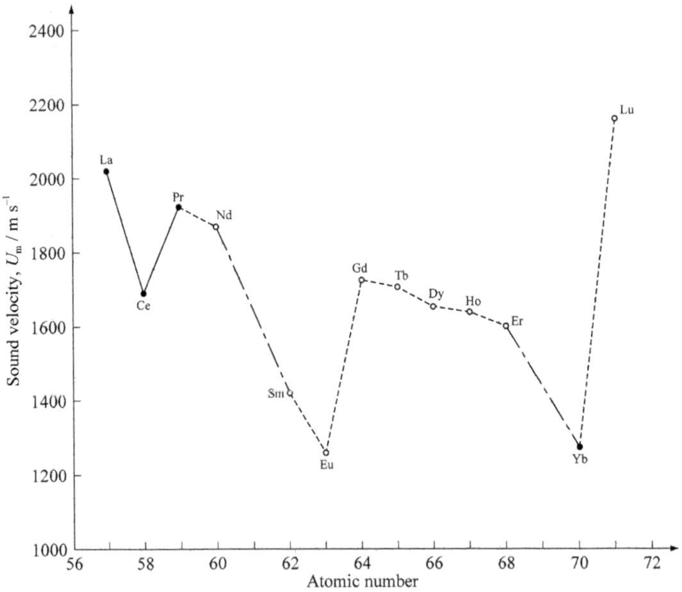

Figure 11.4 *Melting point sound velocity in liquid lanthanoid metals, plotted against atomic number. ●, experimental values; ○, the authors' predicted values.*

properties or physical quantities have also been observed to exhibit analogous periodic variations for the liquid lanthanoid metals [5] (cf. Figures 3.6, 10.2, and 10.3).

11.3 Dimensionless Common Parameters $\xi_T^{1/2}$ and $\xi_E^{1/2}$

11.3.1 Determination of the Common Parameters

We next determine values of the common parameters $\xi_T^{1/2}$ and $\xi_E^{1/2}$ for better predictions of several important thermophysical properties of liquid metallic elements. These parameters are given by the following equations (see Chapter 5).

$$\xi_T^{1/2} = \frac{1}{9.197}\left(\frac{M}{RT_m}\right)^{1/2} U_m \qquad (5.32b)^2$$

$$\xi_E^{1/2} = \left(\frac{M}{2\Delta_l^g H_b}\right)^{1/2} U_m \qquad (5.26b)$$

where M is the molar mass (kg mol^{-1}), T_m is the melting point temperature in K, U_m is the melting point sound velocity (m s^{-1}), R is the molar gas constant (R = 8.314 J mol^{-1} K^{-1}), and $\Delta_l^g H_b$ is the molar evaporation enthalpy at the boiling point temperature (J mol^{-1}).

In Table 11.4, calculated values of $\xi_T^{1/2}$ and $\xi_E^{1/2}$ are also listed; the data used for calculating common parameters are given in Tables 11.4 and 11.5.

In Chapter 5, the present authors pointed out that both the values of $\xi_T^{1/2}$ and $\xi_E^{1/2}$ vary periodically with atomic number. Thus, the values of $\xi_T^{1/2}$ and $\xi_E^{1/2}$ for a large number of liquid metallic elements are plotted against atomic number in Figures 11.5 and 11.6, respectively. As expected, these figures show that both the values of $\xi_T^{1/2}$ and $\xi_E^{1/2}$ vary periodically with atomic number. As can be seen from Figure 11.5, the periodic Group IIIA metals (excluding thallium) and the Group IVB metals occupy the major peaks and the minor peaks, respectively, whereas the Group VIA elements occupy the valleys in the plot. Meanwhile, as shown in Figure 11.6, the Group IIB metals and the Group IIA metals (excluding barium) occupy the major peaks and the minor peaks, respectively, whereas the metallic elements showing anomalous temperature dependence occupy the valleys (silicon, selenium, and cerium, excluding antimony; tellurium (atomic number = 52; $\xi_E^{1/2}$ = 0.602, see Table 13.3) may also occupy the valley) or have small $\xi_E^{1/2}$ values (germanium and plutonium) in the plot. Incidentally, in both figures, mercury lies on a lofty peak.

In general, a periodic variation in a physical quantity or any physical parameter with atomic number should provide an important key for developing rigorous studies on the thermophysical properties of liquid metallic elements. Intuitively, we first can infer,

[2] As noted earlier, the value of the numerical factor of 9.197 (dimensionless) was determined so as to give the minimum S (relative standard deviation) value for the experimentally derived sound velocities of 41 liquid metallic elements which are presently available.

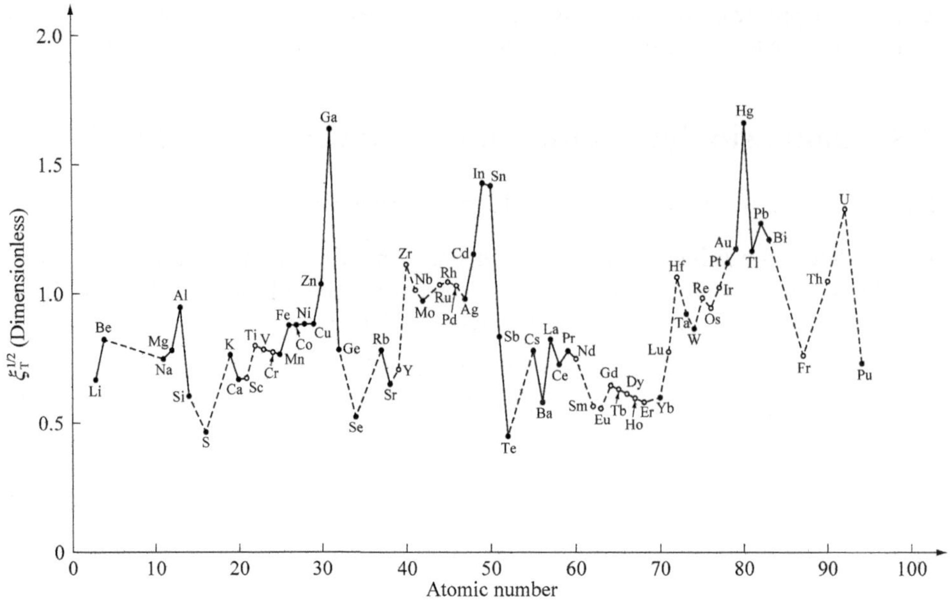

Figure 11.5 *The common parameter, or the dimensionless number, $\xi_T^{1/2}$ for 68 liquid elements.* ●, *determined from experimental, and,* ○, *predicted sound velocity values (see Tables 5.5 and 11.4).*

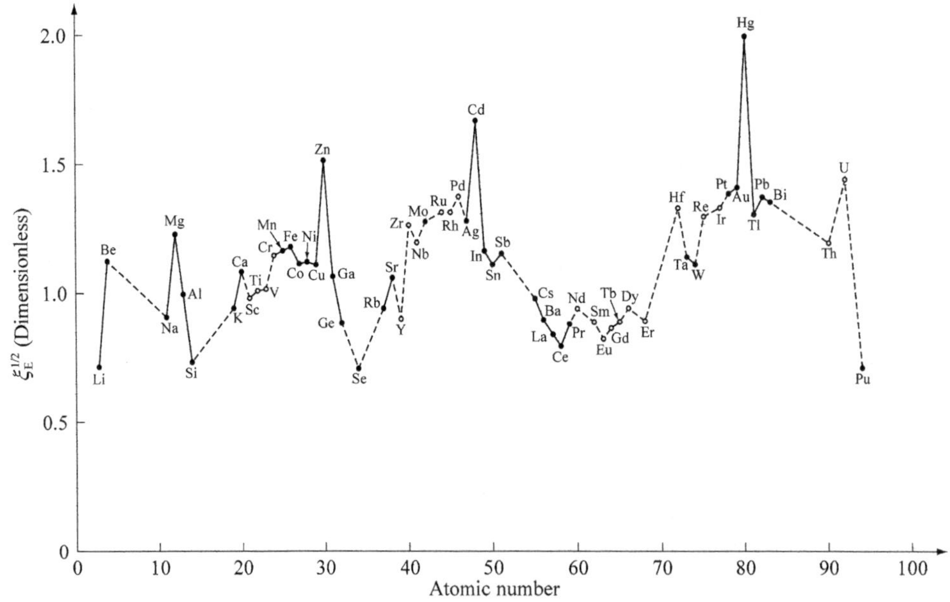

Figure 11.6 *The common parameter, or the dimensionless number, $\xi_E^{1/2}$ for 61 liquid metallic elements.* ●, *determined from experimental, and,* ○, *predicted sound velocity values (see Tables 5.5 and 11.4).*

from the relationships of Figures 11.5 and 11.6, that both the common parameters $\xi_{\mathrm{T}}^{1/2}$ and $\xi_{\mathrm{E}}^{1/2}$ might be relevant to the electronic states, or electron configurations, of the atomic systems. Unfortunately, at present, it is impossible (in practice) to present any rigorous formula based on a quantum mechanical theory (or electron theory) that gives both 'accuracy' and 'universality' to this subject. However, the periodic variations in the values of some physical quantity or parameter (e.g. the Andrade coefficient C_{A}, the common parameters $\xi_{\mathrm{T}}^{1/2}$ and $\xi_{\mathrm{E}}^{1/2}$) with atomic number do allow us to predict (through interpolation) values of the physical quantity or the parameter for those not yet available. Generally speaking, an approach, combined with plural relationships obtained from different angles, is an effective method for predicting unmeasured thermophysical properties or unknown physical quantities. Some examples are given in the subsequent chapters. This expedient approach is useful from the standpoint of materials process science. A few relationships between the common parameters and basic physical quantities have been obtained from empirical or mathematical treatment to date.

11.3.2 Relationships between the Common Parameter $\xi_{\mathrm{T}}^{1/2}$ or $\xi_{\mathrm{E}}^{1/2}$ and Other Physical Quantities, and the Physical Significance of the Common Parameters

(i) On the basis of an empirical consideration, simple linear relations were suggested between the common parameter $\xi_{\mathrm{T}}^{1/2}$ or $\xi_{\mathrm{E}}^{1/2}$ and the repulsive exponent n of the following pair potential of condensed systems.

$$\phi(r) = \varepsilon\left(\frac{a}{r}\right)^n - b\delta^3 \exp(-\delta r), \quad (\varepsilon > 0,\ a > 0,\ n > 3,\ b \geq 0) \qquad (5.27)$$

The common parameters $\xi_{\mathrm{T}}^{1/2}$ and $\xi_{\mathrm{E}}^{1/2}$ are expressed, respectively, in terms of n.

$$\xi_{\mathrm{T}}^{1/2} = 0.0534n + 0.448 \qquad (5.33)$$

$$\xi_{\mathrm{E}}^{1/2} = 0.0707n + 0.543 \qquad (5.28)$$

The dimensionless common parameters, or dimensionless numbers, give an indication of an atom's hardness or softness.

(ii) The repulsive exponent n can be connected with the Grüneisen constant $\gamma_{\mathrm{G,T}}$, or $\gamma_{\mathrm{G,E}}$, in the form

$$n = 6\gamma_{\mathrm{G.E}} - 2 \qquad (5.29a)$$

$$n = 6\gamma_{\mathrm{G.T}} - 2 \qquad (5.29b)$$

Thus,

$$\xi_{\mathrm{T}}^{1/2} = 0.320\gamma_{\mathrm{G,T}} + 0.341 \qquad (5.34)$$

$$\xi_{\mathrm{E}}^{1/2} = 0.424\gamma_{\mathrm{G,E}} + 0.402 \qquad (5.30)$$

The common parameters are also useful in discussing anharmonic effects of atomic motions in liquid metallic elements. For example, the volume (or thermal) expansivity of condensed matter can be calculated on the basis of the Grüneisen relation.

(iii) Combination of Eqs. (5.26a) and (5.32a) (or Eqs. (5.26b) and (5.32b)), gives

$$\Delta_{\mathrm{l}}^{\mathrm{g}} H_b = 351.6\xi T_{\mathrm{m}}, \quad (\xi \equiv \xi_{\mathrm{T}} / \xi_{\mathrm{E}}) \tag{4.10}$$

The values of ξ vary periodically with atomic number; we can predict values for the evaporation enthalpy $\Delta_{\mathrm{l}}^{\mathrm{g}} H_{\mathrm{b}}$ of liquid metallic elements using their melting point temperature data (see Chapter 13).

(iv) The common parameters are related to the ratio of the isobaric and isochoric heat capacities $\gamma_{\mathrm{h}}(\equiv C_P / C_V)$, and the long wavelength of the structure factor $S(0)$:

$$\xi_{\mathrm{T}} = \frac{1}{9.197^2}\left\{\frac{\gamma_{\mathrm{h}}}{S(0)}\right\} = 1.182 \times 10^{-2}\left\{\frac{\gamma_{\mathrm{h}}}{S(0)}\right\} \tag{5.36}$$

$$\xi_{\mathrm{E}} = \frac{R}{2}\frac{T_{\mathrm{m}}}{\Delta_{\mathrm{l}}^{\mathrm{g}} H_{\mathrm{b}}}\left\{\frac{\gamma_{\mathrm{h}}}{S(0)}\right\} = 4.157\frac{T_{\mathrm{m}}}{\Delta_{\mathrm{l}}^{\mathrm{g}} H_{\mathrm{b}}}\left\{\frac{\gamma_{\mathrm{h}}}{S(0)}\right\} \tag{5.37}$$

The common parameters have a close relationship with the structure of metallic liquids.

(v) The common parameters can also be expressed in terms of the isothermal compressibility κ_T, as follows

$$\xi_{\mathrm{T}} = 1.422 \times 10^{-3} \left.\frac{\gamma_{\mathrm{h}} V}{T\kappa_T}\right|_{T=T_{\mathrm{m}}} \tag{5.38}$$

and

$$\xi_{\mathrm{E}} = \left.\frac{\gamma_{\mathrm{h}} V}{2\Delta_l^g H_{\mathrm{b}} \kappa_T}\right|_{T=T_{\mathrm{m}}} \tag{5.39}$$

where V is the molar volume. Using Eq. (5.38), or Eq. (5.39), values of the isothermal compressibility of liquid metallic elements can readily be calculated, provided that their γ_{h} values are available (γ_{h} values are also easily calculable on the basis of Eq. (12.3)); all of the other physical quantities are available for a large number of metallic elements (see Appendix 9).

Incidentally, the numerical factors for Eqs. (5.28), (5.30), (5.33), (5.34), and (5.36) through (5.39) were determined for liquid metallic elements at their melting point temperatures.

At the very end of this chapter, we refer, cursorily, to the temperature coefficient, $(\partial U / \partial T)_P$, of the sound velocity in metallic elements.

Experimentally derived sound velocity data for liquid metallic elements can usually be expressed by the following equation.

$$U = U_m + \left(\frac{\partial U}{\partial T}\right)_P (T - T_m) \tag{11.9}$$

in which the temperature coefficient $(\partial U / \partial T)_P$ is negative. Exceptionally, as already mentioned, six liquid metallic elements are found to display positive values for $(\partial U / \partial T)_P$, over limited ranges of temperatures from their melting points.

Blairs [1] has presented a comparison of experimental and calculated values of $(\partial U / \partial T)_P$, based on five model theories, for 41 liquid metallic elements. The comparison of experimental and calculated results suggest that it is still too soon to make accurate predictions of $(\partial U / \partial T)_P$ for liquid metallic elements, from the point of view of materials process science. To begin with, fairly large discrepancies exist among experimental values for the temperature coefficients of liquid metallic elements; most of the experimental data for the temperature coefficient of liquid metallic elements vary by ± 15 to ± 40 per cent around the mean. There is a need for models which will predict values of the temperature coefficient of liquid metallic elements to an accuracy of ± 10 per cent, although accurate experimental determination of $(\partial U / \partial T)_P$ for metallic liquids is no easy matter.

The velocity of sound provides many pieces of useful information, not only on thermodynamic properties but also on transport properties. Thus, the velocity of sound gives a reasonable connection between thermodynamic properties and transport properties. The following is only one example. Although the Fowler–Born–Green relation connecting surface tension (i.e. a thermodynamic property) and viscosity (i.e. transport property) is an approximate formula, the modified Fowler–Born–Green relation, expressed in terms of the common parameter $\xi_T^{1/2}$, revealed through data for the velocity of sound, or the relationship between surface tension, viscosity and sound velocity, represented by Eq. (6.37) or (15.1), is a more accurate, or an accurate formula (e.g. Tables 15.3 and 15.5).

Moreover, the velocity of sound reflects the repulsive and attractive potential energy of atoms; this leads to an understanding of anharmonic effects of atomic motions.

REFERENCES

1. S. Blairs, *Int. Mater. Rev.*, 52 (2007), 321.
2. S. Blairs, *Phys. Chem. Liq.*, **45** (2007), 399.
3. S. Blairs, *J. Colloid. Interface Sci.*, **302** (2006), 312.
4. I. Yokoyama, S. Naito, and Y. Waseda, *J. Less-Common Met.*, **136** (1987), 25.
5. V.I. Kononenko, A.L. Sukhman, S.L. Gruverman, and V.V. Torokin, *Phys. Stat. Sol. (a)*, **84** (1984), 423.

12

Predictive Models for Volume Expansivity

12.1 Introduction

A thorough knowledge of the mass density[1] (or the number density[2]) and its temperature dependence is indispensable for a clear understanding of all the properties and behaviour of metallic liquids. From a practical point of view, for example, the density and its temperature dependence, or volume expansivity, of a metallic liquid is of great importance in discussing fluid dynamics, heat and mass transfer, and solidification (e.g. crystal growth), since values for its density and volume expansivity are involved in calculations of dimensionless numbers, such as Grashof, Prandtl, Rayleigh, Reynolds, and Schmidt numbers, used in quantifying these various phenomena. Nevertheless, experimentally obtained data for the temperature dependence of the density, or volume expansivity, for practically all liquid metallic elements, have yet to be established. Moreover, studies on predictive models for accurate and reliable values of a metallic liquid's density and volume expansivity have hardly been made from any point of view (e.g. either theoretical, empirical, or through mathematical modelling). Only the Steinberg model (see Chapter 3) for the temperature dependence of the density of liquid metallic elements has been identified. Accordingly, the authors, in Chapter 3, have proposed two more models for the volume expansivity of liquid metallic elements, by combining the Grüneisen relation with the common parameters, ξ_T and ξ_E, which characterize metallic atoms in the liquid state.

These three models are expressed in terms of well-known, or readily available, physical quantities, i.e. the melting and boiling point temperatures, the common parameters, the evaporation enthalpy, and the constant-pressure heat capacity. As shown in Tables 3.3 and 3.4, these three models can reproduce comparatively accurate experimental values for the volume expansivity of the 16 liquid metallic elements selected,

[1] The mass per unit volume of a substance; in SI units it is measured in kg m^{-3}.

[2] The number of particles, atoms, molecules, etc. per unit volume. For example, the number density, denoted by n_0, where $n_0 = N_A / V = (N_A / M)\rho$, in which N_A is Avogadro constant, $V(= M/\rho)$ is the molar volume, M is the molar mass, ρ is the mass density, or simply 'density'.

The Thermophysical Properties of Metallic Liquids: Volume 2 – Predictive Models. First Edition.

Published in 2015 by Oxford University Press.

with good agreement. As such, it can safely be said that these three models are endowed with the necessary conditions of being predictive in both accuracy and universality.

In this chapter, calculated volume expansivities using the Steinberg model and authors' models are compared with experimental data for a large number of liquid metallic elements.

12.2 Predictive Models for Liquid Metallic Element Volume Expansivity

12.2.1 The Steinberg Model

The Steinberg model for the temperature dependence of the density, $\Lambda[\equiv (\partial\rho / \partial T)_P]$, of liquid metallic elements is expressed in the following form [1]:

$$\frac{\Lambda T_{\mathrm{b}}}{D_{00}} = -\,0.23 \tag{3.11}$$

$$D_{00} \equiv \rho_{\mathrm{m}} - \Lambda T_{\mathrm{m}}$$

where subscripts b and m denote boiling and melting points, respectively, and D_{00} (a scale factor) is the density determined by extrapolating from ρ_{m} to absolute zero. It is not necessarily equal to the true density at 0 K.

According to Steinberg's research article [1] published in 1974, the numerical factor of 0.23 appearing in Eq. (3.11) was obtained so as to give the best fit to experimental data (excluding semimetals and Groups VIII and IIB metals).

By rewriting Eq. (3.11), we have an equation for the volume expansivity, $\alpha_{\mathrm{m}}(\equiv -\Lambda / \rho_{\mathrm{m}})$, of liquid metallic elements at their melting point temperatures:

$$\alpha_{\mathrm{m}} = \frac{0.23}{T_{\mathrm{b}} - 0.23\,T_{\mathrm{m}}} \tag{3.12}$$

12.2.2 The Authors' Models

The authors' models for the volume expansivity of liquid metallic elements are based on the Grüneisen relation, which is given by (see Chapters 3 and 5)

$$\alpha = \frac{\gamma_{\mathrm{G}} \kappa_T C_V}{V} \tag{3.13}$$

where γ_{G} is the Grüneisen constant, κ_T is the isothermal compressibility, C_V is the constant-volume heat capacity, and V is the volume.

The Grüneisen constant can be computed using simple empirical equations, as given below:

$$\gamma_{G,T} = 3.12\xi_T^{1/2} - 1.06, \qquad (\xi_T^{1/2} > 0.608) \tag{5.35}$$

$$\gamma_{G,E} = 2.36\xi_E^{1/2} - 0.947, \qquad (\xi_E^{1/2} > 0.755) \tag{5.31}$$

where $\gamma_{G,T}$ and $\gamma_{G,E}$ are the Grüneisen constants calculated from Eq. (5.35) expressed in terms of $\xi_T^{1/2}$, and from Eq. (5.31) in terms of $\xi_E^{1/2}$, respectively.

Substitution of Eq. (5.35), or Eq. (5.31), into Eq. (3.13) leads to:

$$\alpha_m = \left.\frac{A_T}{\xi_T^{1/2} T_m}\right|_{T=T_m} \tag{3.14}$$

$$A_T \equiv \left(0.0368_9 - \frac{0.0125_3}{\xi_T^{1/2}}\right)\frac{C_P}{R} = \left(4.44 - \frac{1.51}{\xi_T^{1/2}}\right) \times 10^{-3} C_P$$

where C_P is the constant-pressure heat capacity and R (= 8.314 J mol^{-1}K^{-1}) is the molar gas constant;

$$\alpha_m = \left.\frac{A_E}{\xi_E^{1/2} \Delta_l^g H_b}\right|_{T=T_m} \tag{3.15}$$

$$A_E \equiv \left(1.18 - \frac{0.474}{\xi_E^{1/2}}\right) C_P$$

where $\Delta_l^g H_b$ is the molar evaporation enthalpy at the boiling point temperature. Table 12.1 lists values of the characteristic common parameters for a large number of liquid metallic elements.

The volume expansivity α_m can also be expressed in the following form. Substitution of Eq. (5.38b), or Eq. (5.39b), into Eq. (3.13) gives

$$\alpha_m = 1.182 \times 10^{-2} \left.\frac{\gamma_{G,T} C_P}{R T \xi_T}\right|_{T=T_m} \tag{12.1}$$

or

$$\alpha_m = \left.\frac{\gamma_{G,E} C_P}{2\Delta_l^g H_b \xi_E}\right|_{T=T_m} \tag{12.2}$$

Values for $\gamma_{G,T}$ and $\gamma_{G,E}$, calculated from Eqs. (5.35) and (5.31), respectively, are given in Table 12.2. Figure 12.1 shows the variation of $\gamma_{G,E} / \xi_E$ with the common parameter ξ_E; a similar variation is obtained for the relationship between $\gamma_{G,T} / \xi_T$, and ξ_T.

Table 12.1 *Values of the dimensionless common parameters which characterize the state of atoms in liquid metallic elements at their melting temperatures.*

Element		ξ_T	ξ_E	$\xi_T^{1/2}$	$\xi_E^{1/2}$
Aluminium	Al	0.900	1.005	0.949	1.002
Antimony	Sb	0.699	1.346	0.836	1.160
Barium	Ba	0.346	0.806	0.588	0.898
Beryllium	Be	0.681	1.270	0.825	1.127
Bismuth	Bi	1.468	1.849	1.211	1.360
Cadmium	Cd	1.346	2.813	1.160	1.677
Caesium	Cs	0.606	0.974	0.778	0.987
Calcium	Ca	0.453	1.185	0.673	1.088
Cerium	Ce	0.533	0.640	0.730	0.800
Chromium	Cr	*0.603**	*1.328*	*0.776*	*1.152*
Cobalt	Co	0.770	1.250	0.878	1.118
Copper	Cu	0.788	1.237	0.887	1.112
Dysprosium	Dy	*0.375*	*0.886*	*0.613*	*0.941*
Erbium	Er	*0.339*	*0.793*	*0.582*	*0.890*
Europium	Eu	*0.312*	*0.682*	*0.558*	*0.826*
Francium	Fr	*0.580*	*0.867***	*0.762*	*0.931***
Gadolinium	Gd	*0.420*	*0.750*	*0.648*	*0.866*
Gallium	Ga	2.702	1.133	1.644	1.064
Germanium	Ge	0.619	0.791	0.787	0.889
Gold	Au	1.381	1.998	1.175	1.414
Hafnium	Hf	*1.151*	*1.776*	*1.073*	*1.333*
Holmium	Ho	*0.361*	*0.836***	*0.601*	*0.914***
Indium	In	2.045	1.367	1.430	1.169
Iridium	Ir	*1.049*	*1.778*	*1.024*	*1.333*
Iron	Fe	0.774	1.392	0.880	1.180
Lanthanum	La	0.678	0.710	0.823	0.843
Lead	Pb	1.627	1.909	1.275	1.382

Table 12.1 *(continued)*

Element		ξ_T	ξ_E	$\xi_T^{1/2}$	$\xi_E^{1/2}$
Lithium	Li	0.451	0.507	0.672	0.712
Lutetium	Lu	*0.601*	*1.043***	*0.775*	*1.021***
Magnesium	Mg	0.619	1.522	0.787	1.234
Manganese	Mn	0.588	1.365	0.767	1.168
Mercury	Hg	2.779	4.003	1.667	2.001
Molybdenum	Mo	0.955	1.648	0.977	1.284
Neodymium	Nd	*0.556*	*0.891*	*0.746*	*0.944*
Nickel	Ni	0.791	1.265	0.889	1.125
Niobium	Nb	*1.029*	*1.430*	*1.015*	*1.196*
Osmium	Os	*0.910*	*1.531***	*0.954*	*1.237***
Palladium	Pd	*1.064*	*1.895*	*1.032*	*1.377*
Platinum	Pt	1.266	1.938	1.125	1.392
Plutonium	Pu	0.543	0.507	0.737	0.712
Potassium	K	0.582	0.889	0.763	0.943
Praseodymium	Pr	0.617	0.784	0.785	0.885
Rhenium	Re	*0.975*	*1.677*	*0.987*	*1.295*
Rhodium	Rh	*1.093*	*1.733*	*1.045*	*1.316*
Rubidium	Rb	0.609	0.889	0.780	0.943
Ruthenium	Ru	*1.073*	*1.731*	*1.036*	*1.316*
Samarium	Sm	*0.321*	*0.792*	*0.567*	*0.890*
Scandium	Sc	*0.459*	*0.959*	*0.677*	*0.980*
Selenium	Se	0.275	0.504	0.525	0.710
Silicon	Si	0.364	0.546	0.604	0.739
Silver	Ag	0.967	1.659	0.983	1.288
Sodium	Na	0.562	0.823	0.750	0.907
Strontium	Sr	0.429	1.140	0.655	1.068
Sulphur	S	0.219	–	0.467	–
Tantalum	Ta	0.853	1.311	0.924	1.145

continued

Table 12.1 *(continued)*

Element		ξ_T	ξ_E	$\xi_T^{1/2}$	$\xi_E^{1/2}$
Tellurium	Te	0.198	0.362**	0.445	0.602**
Terbium	Tb	*0.406*	*0.792*	*0.637*	*0.890*
Thallium	Tl	1.371	1.717	1.171	1.310
Thorium	Th	*1.098*	*1.435*	*1.048*	*1.198*
Tin	Sn	2.029	1.238	1.424	1.113
Titanium	Ti	*0.640*	*1.015*	*0.800*	*1.008*
Tungsten	W	0.761	1.235	0.872	1.111
Uranium	U	*1.776*	*2.079*	*1.333*	*1.442*
Vanadium	V	*0.618*	*1.036*	*0.786*	*1.018*
Ytterbium	Yb	0.366	1.025**	0.605	1.012**
Yttrium	Y	*0.505*	*0.811*	*0.710*	*0.900*
Zinc	Zn	1.090	2.309	1.044	1.520
Zirconium	Zr	*1.255*	*1.613*	*1.120*	*1.270*

* Numerical figures in italics denote values determined on the basis of sound velocities predicted using the authors' models (see Chapter 11).

** Determined on the basis of boiling point evaporation enthalpies predicted using the authors' models (see Chapter 13).

Table 12.2 *Values of the Grüneisen constants $\gamma_{G,T}$ and $\gamma_{G,E}$ of liquid metallic elements at their melting point temperatures.*

Element		$\gamma_{G,T}$	$\gamma_{G,E}$
Aluminium	Al	1.90	1.42
Antimony	Sb	1.55	1.79
Barium	Ba	0.775	1.17
Beryllium	Be	1.51	1.71
Bismuth	Bi	2.72	2.26
Cadmium	Cd	2.56	3.01
Caesium	Cs	1.37	1.38
Calcium	Ca	1.04	1.62

Table 12.2 *(continued)*

Element		$\gamma_{G,T}$	$\gamma_{G,E}$
Cerium	Ce	1.22	0.941
Chromium	Cr	*1.36**	*1.77*
Cobalt	Co	1.68	1.69
Copper	Cu	1.71	1.68
Dysprosium	Dy	*0.853*	*1.27*
Erbium	Er	*0.756*	*1.15*
Europium	Eu	*0.681*	*1.00*
Francium	Fr	*1.32*	*1.25***
Gadolinium	Gd	*0.962*	*1.10*
Gallium	Ga	4.07	1.56
Germanium	Ge	1.40	1.15
Gold	Au	2.61	2.39
Hafnium	Hf	*2.29*	*2.20*
Holmium	Ho	*0.815*	*1.21***
Indium	In	3.40	1.81
Iridium	Ir	*2.13*	*2.20*
Iron	Fe	1.68	1.84
Lanthanum	La	1.51	1.04
Lead	Pb	2.92	2.31
Lithium	Li	1.04	0.733
Lutetium	Lu	*1.36*	*1.46***
Magnesium	Mg	1.40	1.97
Manganese	Mn	1.33	1.81
Mercury	Hg	4.14	3.78
Molybdenum	Mo	1.99	2.08
Neodymium	Nd	*1.27*	*1.28*
Nickel	Ni	1.71	1.71

continued

Table 12.2 *(continued)*

Element		$\gamma_{G,T}$	$\gamma_{G,E}$
Niobium	Nb	*2.11*	*1.88*
Osmium	Os	*1.92*	*1.97***
Palladium	Pd	*2.16*	*2.30*
Platinum	Pt	2.45	2.34
Plutonium	Pu	1.24	0.733
Potassium	K	1.32	1.28
Praseodymium	Pr	1.39	1.14
Rhenium	Re	*2.02*	*2.11*
Rhodium	Rh	*2.20*	*2.16*
Rubidium	Rb	1.37	1.28
Ruthenium	Ru	*2.17*	*2.16*
Samarium	Sm	*0.709*	*1.15*
Scandium	Sc	*1.05*	*1.37*
Selenium	Se	0.578	0.729
Silicon	Si	0.824	0.797
Silver	Ag	2.01	2.09
Sodium	Na	1.28	1.19
Strontium	Sr	0.984	1.57
Tantalum	Ta	1.82	1.76
Tellurium	Te	0.328	0.476**
Terbium	Tb	*0.927*	*1.15*
Thallium	Tl	2.59	2.14
Thorium	Th	*2.21*	*1.88*
Tin	Sn	3.38	1.68
Titanium	Ti	*1.44*	*1.43*
Tungsten	W	1.66	1.67
Uranium	U	*3.10*	*2.46*

Table 12.2 *(continued)*

Element		$\gamma_{G,T}$	$\gamma_{G,E}$
Vanadium	V	*1.39*	*1.46*
Ytterbium	Yb	0.828	1.44**
Yttrium	Y	*1.16*	*1.18*
Zinc	Zn	2.20	2.64
Zirconium	Zr	*2.43*	*2.05*

* Numerical figures in italics denote values determined on the basis of sound velocities predicted from the authors' models (see Chapter 11).
** Determined on the basis of boiling point evaporation enthalpies predicted from the authors' models (see Chapter 13).

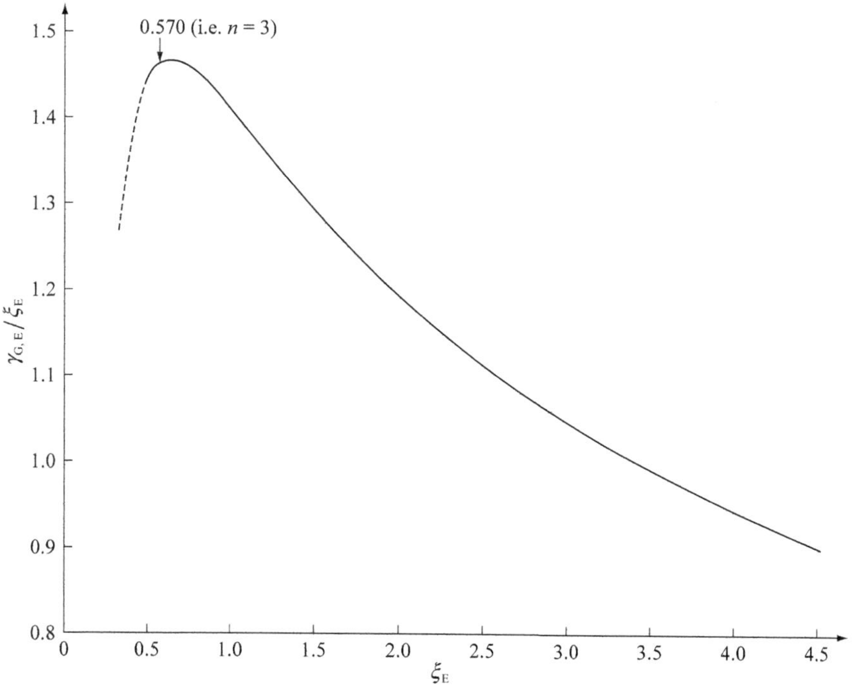

Figure 12.1 *Relationship between $\gamma_{G,E}$ / ξ_E and ξ_E for liquid metallic elements at their melting point temperatures.*

Incidentally, we can now obtain a simple equation for calculating the ratio of the heat capacities γ_h: by combining Eqs. (4.10), (5.1), (5.26), and (12.2)), Eq. (5.3) can be written, at the melting point, as

$$\begin{aligned}\gamma_h \equiv \frac{C_P}{C_V} &= 1 + \left.\frac{\alpha^2 VT}{\kappa_S C_P}\right|_{T=T_m} \\ &= 1 + \frac{1}{9.197^2}\left.\frac{\gamma_{G,E}^2 C_P}{R\xi_T}\right|_{T=T_m} \\ &= 1 + 1.422\times10^{-3}\left.\frac{\gamma_{G,E}^2 C_P}{\xi_T}\right|_{T=T_m}\end{aligned} \tag{12.3}$$

Values for the ratio of the heat capacities γ_h calculated from Eq. (12.3) are given in Appendix 10, together with those calculated by Kleppa [2] using thermodynamic relations.

12.3 Comparison with Experiment

Experimentally derived data for the volume expansivity or the temperature dependence of density have been reported for a large number of liquid metallic elements. At present, experimental volume expansivity data are available for some 66 liquid metallic elements. Unfortunately, there are scarcely any accurate and reliable data for the volume expansivity of liquid metallic elements. Large discrepancies can exist among experimental volume expansivities that have been obtained for a considerable number of liquid metallic elements. For example, Table 12.3 gives reported volume expansivity data for several liquid metals; they show a spread of approximately ± 20 to ± 50 per cent, and these results reflect the difficulties associated with accurate measurements on the volume expansivity of metallic liquids. Mercury may be the only liquid metal whose volume expansivity is accurately known. Unfortunately, mercury is an anomalous metal from a viewpoint of liquid state physics. This must be uppermost in our thoughts.

Values for the volume expansivity of 22 liquid metallic elements for the *s*-block, the *p*-block metals (including semimetals), and zinc group (i.e. Group IIB) metals were calculated from Eqs. (3.9), (3.12), (12.1), or (3.14), and (12.2), or (3.15). As mentioned previously, experimental data for the volume expansivity of almost all liquid metallic elements have yet to be established; so that experimental data have been taken from two specialist works [3,4]. Calculated volume expansivity values and the corresponding δ_i, Δ, and S values are shown in Tables 12.4 to 12.6, respectively. As can be seen from these tables, Eqs. (3.9), (3.12), (12.1), and (12.2), on average (i.e. in view of Δ and S values), provide approximately the same results for the experimental data taken from the two specialist works, although somewhat or fairly different experimental values have been adopted in them. Incidentally, Eq. (12.1), or (3.14), can be regarded as a modified model for Eq. (3.9). It can safely be said that the three models represented by Eqs. (3.12), (12.1), and (12.2) perform well with the 22 liquid metallic elements,

Table 12.3 *Discrepancy in experimental values for the volume expansivity of several liquid metals at their melting point temperatures.*

Metal		$(\alpha_m)_{exp}$ † $10^{-4}K^{-1}$	Published year	Spread‡ %
Copper	Cu	0.968	2006	± 39
		1.1	1996	
		1.0††	1974	
		0.836	1972	
		0.98	1971	
		1.2	1963	
		1.9	1929	
Nickel	Ni	0.82	2004	± 38
		1.28	2004	
		0.85	1996	
		1.51††	1974	
		1.83	1972	
		1.38	1969	
		1.1	1967	
		1.3	1967	
Platinum	Pt	0.50	2006	± 50
		0.88	1993	
		1.2	1991	
		0.68	1979	
		1.52††	1974	
Silver	Ag	0.81	2006	± 21
		1.05	1975	
		0.98††	1974	
		0.97	1951	
		1.08	1937	
		0.70	1929	
Titanium	Ti	0.53	2009	± 39
		1.21	2000	
		0.56††	1974	
		0.547	1969	

† Sources of data are given in Table 12.8.
‡ A spread of experimentally derived values.
†† A preferred value (for α_m) which Steinberg [1] considered reasonable.

Table 12.4 *Comparison of experimental values for the liquid volume expansivity of the s-block, the p-block metals (including semimetals), and the zinc group metals at their melting point temperatures with those calculated from four equations.*

Metal		$(\alpha_m)_{exp}$† / 10^{-4} K^{-1}		$(\alpha_m)_{cal}$ / 10^{-4} K^{-1}				C_P##
			range‡	Eq. (3.9)††	Eq. (3.12)	Eq. (12.1)	Eq. (12.2)	J mol^{-1} K^{-1}
Aluminium	Al	1.5	1.0–1.7	0.96	0.89	1.02	0.765	31.8
Antimony	Sb	1.3	0.69–1.82	1.0	1.4	1.10	1.27	31.4
Barium	Ba	0.81	–	0.90	1.2	0.997	1.50	31.3
Beryllium	Be	0.71	–	0.58	0.96	0.635	0.719	31.4
Bismuth	Bi	1.17	1.07–1.40	1.7	1.3	1.48	1.23	30.5
Cadmium	Cd	1.52	1.42–1.76	1.5	2.5	1.35	1.59	29.7
Caesium	Cs	3.1	3.0–3.3	3.0	2.6	3.95	3.98	37
Calcium	Ca	1.6	–	0.81	1.5	0.980	1.53	33.47
Francium	Fr	3.37	–	3.0	2.6	3.41	3.24	31.7
Gallium	Ga	0.92	0.84–1.2	3.0	0.96	1.96	0.751	27.7
Indium	In	0.97	0.97–1.3	2.1	1.0	1.63	0.870	29.7
Lead	Pb	1.24	1.12–1.25	1.5	1.2	1.34	1.06	31.5
Lithium	Li	1.9	–	2.0	1.5	2.19	1.54	30.33
Magnesium	Mg	1.6	–	0.98	2.0	1.12	1.57	32.1
Mercury	Hg	1.78	–	3.8	4.0	2.58	2.35	28.5
Potassium	K	2.9	2.7–3.0	2.7	2.4	3.08	2.99	32.1
Rubidium	Rb	3.0	2.8–3.2	2.9	2.6	3.49	3.25	34.0
Sodium	Na	2.54	2.43–2.64	2.4	2.1	2.78	2.59	31.86
Strontium	Sr	1.1	–	0.86	1.6	0.963	1.53	31.0
Thallium	Tl	1.15	1.03–1.27	1.6	1.4	1.42	1.17	30.5
Tin	Sn	0.87	0.77–1.0	1.8	0.83	1.39	0.692	29.7
Zinc	Zn	1.5	1.4–1.69	1.3	2.3	1.30	1.56	31.3

† From Gale and Tolemeier [3] (cf.also Steinberg [1]);
‡ See Steinberg [1];
†† $\alpha_m = 0.09 / T_m$.
Data are given in Chapter 17.

Table 12.5 *Parameter δ_i, Δ, and S values needed for statistical assessment, obtained from four equations.*

Metal		δ_i / %			
		Eq. (3.9)	Eq. (3.12)	Eq. (12.1)	Eq. (12.2)
Aluminium	Al	56	69	47	96
Antimony	Sb	30	−7.1	18	2.4
Barium	Ba	−10	−33	−19	−46
Beryllium	Be	22	−26	12	−1.3
Bismuth	Bi	−31	−10	−20.9	−4.9
Cadmium	Cd	1.3	−39	12.6	−4.4
Caesium	Cs	3.3	19	−22	−22
Calcium	Ca	98	6.7	63	4.6
Francium	Fr	12	30	−1.2	4.0
Gallium	Ga	−69	−4.2	−53	23
Indium	In	−54	−3.0	−40	11
Lead	Pb	−17	3.3	−7.5	17
Lithium	Li	−5.0	27	−13	23
Magnesium	Mg	63	−20	43	1.9
Mercury	Hg	−53	−56	−31.0	−24.3
Potassium	K	7.4	21	−5.8	−3.0
Rubidium	Rb	3.4	15	−14	−7.7
Sodium	Na	5.8	21	−8.6	−1.9
Strontium	Sr	28	−31	14	−28
Thallium	Tl	−28	−18	−19.0	−1.7
Tin	Sn	−52	4.8	−37	26
Zinc	Zn	15	−35	15	−3.8
	Δ(22) %	30.2	22.7	23.5	16.3
	S(22)	0.397	0.282	0.286	0.266

Table 12.6 *Comparison of experimental values for the liquid volume expansivity of the s-block, the p-block metals (including semimetals), and the zinc group metals at their melting point temperatures with those calculated from four equations, together with δ_i, Δ, and S values.*

Element		$(\alpha_m)_{exp}$† 10^{-4}K^{-1}	$(\alpha_m)_{cal}$ / 10^{-4} K^{-1}				δ_i / %			
			Eq. (3.9)	Eq. (3.12)	Eq. (12.1)	Eq. (12.2)	Eq. (3.9)	Eq. (3.12)	Eq. (12.1)	Eq. (12.2)
Aluminium	Al	1.2	0.96	0.89	1.02	0.765	25	35	18	57
Antimony	Sb	0.872	1.0	1.4	1.10	1.27	−13	−38	−20.7	−31.3
Barium	Ba	1.58	0.90	1.2	0.997	1.50	76	32	58.5	5.3
Beryllium	Be	0.688	0.58	0.96	0.635	0.719	19	−28	8.3	−4.3
Bismuth	Bi	1.32	1.7	1.3	1.48	1.23	−22	1.5	−10.8	7.3
Cadmium	Cd	1.45	1.5	2.5	1.35	1.59	−3.3	−42	7.4	−8.8
Caesium	Cs	3.44	3.0	2.6	3.95	3.98	15	32	−12.9	−13.6
Calcium	Ca	1.62	0.81	1.5	0.980	1.53	100	8.0	65.3	5.9
Francium‡	Fr	3.37	3.0	2.6	3.41	3.24	12	30	−1.2	4.0
Gallium	Ga	0.99	3.0	0.96	1.96	0.751	−67	3.1	−49	32
Indium	In	0.968	2.1	1.0	1.63	0.870	−54	−3.2	−40.6	11.3
Lead	Pb	1.23	1.5	1.2	1.34	1.06	−18	2.5	−8.2	16.0
Lithium	Li	1.9	2.0	1.5	2.19	1.54	−5	27	−13	23
Magnesium	Mg	1.66	0.98	2.0	1.12	1.57	69	−17	48.2	5.7
Mercury	Hg	1.78	3.8	4.0	2.58	2.35	−53	−56	−31.0	−24.3
Potassium	K	2.76	2.7	2.4	3.08	2.99	2.2	15	−10.4	−7.7
Rubidium	Rb	3.38	2.9	2.6	3.49	3.25	17	30	−3.2	4.0
Sodium	Na	2.55	2.4	2.1	2.78	2.59	6.3	21	−8.3	−1.5
Strontium	Sr	1.1‡	0.86	1.6	0.963	1.53	28	−31	14	−28
Thallium	Tl	1.27	1.6	1.4	1.42	1.17	−21	−9.3	−10.6	8.5
Tin	Sn	0.875	1.8	0.83	1.39	0.692	−51	5.4	−37.1	26.4
Zinc	Zn	1.67	1.3	2.3	1.30	1.56	28	−27	28.5	7.1
						Δ(22) %	32.0	22.5	23.0	15.1
						S(22)	0.416	0.268	0.295	0.201

† From reference [4].
‡ From Gale and Tolemeier [3].

Table 12.7 *Comparison of experimental values of the liquid volume expansivity of (a) rare earth metals and (b) semiconductors at their melting point temperatures with those calculated from Eqs. (3.12), (12.1), and (12.2).*

(a)

Metal		$(\alpha_m)_{exp}$† $10^{-4}K^{-1}$	$(\alpha_m)_{cal}/10^{-4}\,K^{-1}$			C_P‡ J $mol^{-1}\,K^{-1}$
			Eq. (3.12)	Eq. (12.1)	Eq. (12.2)	
Cerium	Ce	1.1	0.66	1.14	0.88	37.66
		0.339		1.06	0.82	35
		0.354		0.96	0.74	31.6
Dysprosium	Dy	1.71	0.96	0.96	1.43	49.92
Erbium	Er	1.77	0.84	0.68	1.04	38.70
Europium	Eu	5.5	1.5	1.08	1.59	38.07
Gadolinium	Gd	1.3	0.72	0.76	0.87	37.15
		0.600		0.69	0.79	33.5
Holmium	Ho	1.71	0.89	0.81	1.20	43.93
		0.51		0.62	0.91	33.5
Lanthanum	La	1.0	0.66	0.91	0.63	34.31
		0.398				
Lutetium	Lu	2.09	0.71	0.80	0.86	47.91
Neodymium	Nd	1.1	0.75	1.22	1.23	48.79
		0.790		0.84	0.85	33.5
				1.10	1.11	44.0
Praseodymium	Pr	1.4	0.65	1.15	0.94	42.97
		0.382		0.89	0.73	33.5
				1.11	0.91	41.57
Samarium	Sm	0.70	1.3	1.17	1.90	50.21
		1.33		0.78	1.27	33.5
Scandium	Sc	2.0	0.85	0.80	1.04	44.35
				0.60	0.78	33.5
Terbium	Tb	1.73	0.74	0.93	1.15	46.44
		0.630				
Ytterbium	Yb	4.84	1.9	1.08	1.89	36.82
		1.12				
Yttrium	Y	0.69	0.72	0.73	0.74	39.79
		0.51††		0.61	0.62	33.5

Table 12.7 *(continued)*

(b)

Semiconductor		$(\alpha_m)_{exp} / 10^{-4}$ K^{-1}			$(\alpha_m)_{cal} / 10^{-4}$ K^{-1}			C_P‡ J mol^{-1} K^{-1}
			Range	**Ref.**	**Eq. (3.12)**	**Eq. (12.1)**	**Eq. (12.2)**	
Germanium	Ge	0.89	0.86–0.89	[1,3]	0.81	0.734	0.603	27.61
		0.666	–	[7]		0.779	0.640	29.3
		0.821	–	[4]				
Selenium	Se	2.93	2.93–3.08	[1,3]	2.7	2.00	2.53	33.15
		3.61	–	[4]		2.12	2.68	35.1
Silicon	Si	1.4	–	[1,3]	0.73	0.518	0.503	27.2
		1.03	–	[4]		0.556	0.540	29.2
		0.78	–	[8]				
Tellurium	Te	1.3	0.90–1.3	[1,3]	2.1	1.23	1.77	37.6
		0.630	–	[4]				

† From Kononenko et al. [5];
‡ Data are given in Chapter 17;
†† From Ishikawa and Paradis [6].

Table 12.8 *Comparison of experimental values for the liquid volume expansivity of the d-block transition metals plus plutonium and uranium at their melting point temperatures with those calculated from Eqs. (3.12), (12.1), and (12.2).*

Metal		$(\alpha_m)_{exp} / 10^{-4}$ K^{-1}		$(\alpha_m)_{cal} / 10^{-4}$ K^{-1}			C_P† J mol^{-1} K^{-1}
			Ref.	**Eq. (3.12)**	**Eq. (12.1)**	**Eq. (12.2)**	
Chromium	Cr	1.1	[1,3]	0.94	0.60	0.79	41
		1.76	[9]		0.747	0.972	50.71
		0.48	[4]				
Cobalt	Co	1.40	[3]	0.82	0.849	0.855	48.45
		(0.92–2.13)‡	[1]		0.61	0.62	35
		1.27	[4]		0.689	0.694	39.3

Table 12.8 *(continued)*

Metal		$(\alpha_m)_{exp}$ / 10^{-4} K^{-1}		$(\alpha_m)_{cal}$ / 10^{-4} K^{-1}			C_P† J mol^{-1} K^{-1}
			Ref.	Eq. (3.12)	Eq. (12.1)	Eq. (12.2)	
Copper	Cu	1.0	[1,3]	0.91	0.717	0.704	31.5
		(0.90–1.25)‡	[1]		0.751	0.737	33.0
		0.968	[10]				
		1.9	[10]				
		1.1	[10]				
		0.98	[10]				
		1.2	[10]				
		0.836	[11]				
Gold	Au	0.632	[10]	0.82	0.589	0.539	29.3
		0.931	[10]				
		0.738	[10]				
Hafnium	Hf	0.47	[6]	0.53	0.378	0.363	33.47
Iridium	Ir	0.44	[6]	0.56	0.444	0.459	41.84
Iron	Fe	1.3	[1,3]	0.85	0.758	0.829	44.4
		(0.88–2.11)‡	[1]		0.785	0.859	46.0
					0.735	0.804	43.05
Manganese	Mn	1.6	[1,3]	1.2	0.974	1.33	46.02
		1.2	[4]				
Molybdenum	Mo	0.66	[6]	0.54	0.413	0.431	40.33
		0.88	[6]		0.351	0.367	34.3
					0.56	0.59	55
Nickel	Ni	1.51	[1,3]	0.82	0.700	0.698	39.3
		(0.86–1.70)‡	[1]		0.648	0.647	36.4
		0.82	[6]		0.768	0.766	43.1
		0.85	[6]				
		1.38	[6]				

Table 12.8 *(continued)*

Metal		$(\alpha_m)_{exp}$ / 10^{-4} K^{-1}		$(\alpha_m)_{cal}$ / 10^{-4} K^{-1}			C_P† J mol^{-1} K^{-1}
			Ref.	Eq. (3.12)	Eq. (12.1)	Eq. (12.2)	
		1.1	[6]				
		1.3	[6]				
		1.83	[6]				
		1.28	[6]				
Niobium	Nb	0.50	[6]	0.52	0.443	0.395	41.8
		0.70	[6]		0.355	0.316	33.5
Osmium	Os	0.607	[6]	0.51	–	–	–
Palladium	Pd	1.17	[1,3]	0.82	0.548	0.584	34.73
		0.72	[6]		0.65	0.69	41
		1.169	[6]				
		1.126	[6]				
		1.207	[4]				
Platinum	Pt	1.52	[1,3]	0.63	0.468	0.447	34.7
		0.50	[6]				
		1.2	[6]				
		0.88	[6]				
		0.68	[6]				
Rhenium	Re	0.42	[6]	0.45	–	–	–
Rhodium	Rh	0.70	[6]	0.67	0.536	0.526	41.84
		0.84	[6]		0.412	0.405	32.2
		0.41	[6]				
Ruthenium	Ru	0.52	[6]	0.60	0.462	0.459	41.84
					0.396	0.394	35.9
Silver	Ag	0.98	[1,3]	1.1	0.802	0.833	33.47
		(0.98–1.0)‡	[1]		0.731	0.759	30.5
		0.81	[10]				
		0.70	[10]				

Table 12.8 *(continued)*

Metal		$(\alpha_m)_{exp}$ / 10^{-4} K^{-1}		$(\alpha_m)_{cal}$ / 10^{-4} K^{-1}			C_P† J mol^{-1} K^{-1}
			Ref.	Eq. (3.12)	Eq. (12.1)	Eq. (12.2)	
		0.97	[10]				
		1.05	[10]				
		1.08	[10]				
Tantalum	Ta	0.58	[6]	0.46	0.385	0.373	41.84
		0.90	[6]				
Titanium	Ti	0.56	[1,3]	0.74	0.690	0.685	41.84
		(0.56–1.7)‡	[1]		0.552	0.548	33.5
		0.53	[6]		0.762	0.756	46.2
		0.547	[6]				
		1.21	[6]				
Tungsten	W	0.657	[6]	0.46	0.446	0.448	53.0
		0.59	[6]				
Vanadium	V	0.60	[1,3]	0.72	0.714	0.749	48.7
		0.90	[12]		0.582	0.611	39.7
		0.53	[12]				
Zirconium	Zr	0.43	[6]	0.55	0.434	0.366	33.5
		0.46	[6]		0.524	0.444	40.7
Plutonium	Pu	0.847	[1,3]	0.70	–	–	–
		(0.847–0.913)‡	[1]		–	–	–
Uranium	U	0.596	[1,3]	0.56	0.857	0.681	48.66
		0.576	[4]		0.675	0.536	38.3
		0.927	[13]				
Thorium	Th	–	–	0.50	0.651	0.554	46.02

† Data are given in Chapter 17;
‡ Range (see Steinberg [1]).

giving $\Delta(22)$ and $S(22)$ values of approximately 15 to 23 per cent and 0.20 to 0.30, respectively, as compared with experimental uncertainties of $\pm$ 10 to $\pm$ 30 per cent, or more. These three models have potential as predictive models for all liquid metallic elements; Eq. (12.2) especially performs very well with the *s*-block, the *p*-block, and zinc group metals. With the exception of several liquid metallic elements, values calculated from Eq. (12.2) fall within the range of uncertainties associated with experimental measurements. Moreover, the model represented by Eq. (12.2) may be worthy of further consideration in the following respects: First, the physical significance of Eq. (12.2) is relatively clear. To elaborate on this, Eq. (12.2) is based on the well-known Grüneisen relation, and further, a relationship between the repulsive exponent n and the common parameter $\xi_E^{1/2}$, i.e. Eq. (5.28), holds well for various metals. In contrast, the numerical factor of 0.23 (the reciprocal of 4.4), appearing in Eq. (3.12), was determined so as to give the best fit to the experimental data (excluding semimetals and Groups VIII and IIB metals) reported before 1973 [1]. Secondly, Eq. (12.2) makes considerable improvements on the Steinberg model for the liquid volume expansivity of Group IIB metals, as well as on the model given by Eq. (12.1) for that of gallium, indium, and tin. The Steinberg model, i.e. Eq. (3.12), also performs very well for the liquid metallic elements shown in Tables 12.4 to 12.6, except for several elements. This is in spite of a simple equation in terms of well-known material constants of only melting and boiling point temperatures. Nevertheless, this does not necessarily guarantee that these three models represented by Eqs. (3.12), (12.1), and (12.2) give accurate and reliable values for the volume expansivity of liquid rare earth and transition metals; therefore, accurate experimental data are needed for those metals' liquid volume expansivities.

The liquid volume expansivities of rare earth metals, the *d*-block transition metals, etc. were calculated from Eqs. (3.12), (12.1), and (12.2). In Table 12.7, the result of calculations for 15 rare earth metals and four semiconductors are compared with experimental values. Similarly, both experimental and calculated values for the volume expansivity of 23 *d*-block transition metals plus actinide metals are given in Table 12.8. Roughly speaking, it appears that the three models represented by Eqs. (3.12), (12.1), and (12.2) provide reasonable values for their liquid volume expansivity.

Unfortunately, at present, detailed discussions about the applicability of these models for the volume expansivity of a great number of liquid metallic elements (e.g. determinations of δ_i, Δ and S values) will not yield fruitful results, owing to large discrepancies, or uncertainties, in their experimental volume expansivities and somewhat smaller discrepancies, or uncertainties, in their experimental constant-pressure heat capacities. Nonetheless, both the volume expansivity α, and the constant-pressure heat capacity C_P, of liquid metallic elements are crucially important thermophysical properties from an academic as well as a technical point of view. For the future, and to determine the true capability of these predictive models, we need to have accurate experimental data (to an accuracy of $\pm$10 per cent or better) for at least 40 liquid metallic elements, including various kinds of metallic elements, for α and C_P.

REFERENCES

1. D.J. Steinberg, *Met. Trans.*, **5** (1974), 1341.
2. O.J. Kleppa, *J. Chem. Phys.*, **18** (1950), 1331.
3. W.F. Gale and T.C. Tolemeier (eds.), *Smithells Metals Reference Book*, 8th ed., Elsevier Butterworth-Heinemann, Oxford, 2004, **14**–10; see also [1].
4. The Japan Institute of Metals, *Metals Data Book*, 4th ed., Maruzen Company, Tokyo, 2004, p.16.
5. V.I. Kononenko, A.L. Sukhman, S.L. Gruverman, and V.V. Torokin, *Phys. Stat. Sol. (a)*, **84** (1984), 423.
6. T. Ishikawa and P.-F. Paradis, in *High-Temperature Measurements of Materials*, H. Fukuyama and Y. Waseda (eds.), Springer, Berlin, 2009, p.182.
7. L. Martin-Garin, M. Gemez, P. Bedon, and P. Desre, *J. Less-Common Met.*, **41** (1975), 65.
8. K. Higuchi, K. Kimura, A. Mizuno, M. Watanabe, Y. Katayama, and K. Kuribayashi, *Meas. Sci. Technol.*, **16** (2005), 381.
9. Y. Tsu, K. Takano, S. Watanabe, and Y. Shiraishi, *Tohoku Daigaku Senko Seiren Kenkyuusho Iho*, **34** (1978), 131.
10. J. Brillo, I. Egry, and I. Ho, *Int. J. Thermophys.*, **27** (2006), 494.
11. S. Watanabe and T. Saito, *Trans. JIM.*, **13** (1972), 186.
12. P.-F. Paradis, T. Ishikawa, T. Aoyama, and S. Yoda, *J. Chem. Thermodyn.*, **34** (2002), 1929.
13. W.G. Rohr and L.J. Wittenberg, *J. Phys. Chem.*, **74** (1970), 1151.

13

Predictions of Evaporation Enthalpy

13.1 Introduction

The enthalpy of evaporation, or the heat of vaporization, of a liquid metallic element is a very useful thermophysical quantity which is directly related to its cohesive (or binding) energy. Thus the enthalpy of evaporation is closely connected with other thermophysical properties of metallic liquids. For example, the surface tension of liquid metallic elements (see the Skapski model), their velocity of sound (see the Gitis–Mikhailov model), their volume expansivity, their liquid range, and so forth, are expressed in terms of the enthalpy of evaporation. A wealth of experimental data for the evaporation enthalpy of liquid metallic elements at their boiling point temperatures has been gathered, and compiled in references (e.g. Table 17.3). However, experimentally derived data for the evaporation enthalpy of some 12 liquid metallic elements still remain unavailable.

In this chapter, the authors predict or calculate values for the evaporation enthalpies of 12 liquid metallic elements, i.e. actinium, arsenic, francium, holmium, lutetium, osmium, promethium, radium, technetium, tellurium, thulium, and ytterbium, at their boiling temperatures. In addition, to check their reliability, relationships between calculated evaporation enthalpies together with those observed experimentally, and other thermophysical properties or quantities, are illustrated with diagrams.

13.2 Predictive Models for Evaporation Enthalpy

13.2.1 Trouton's Rule

Trouton's rule can be expressed in the form (see Subsection 1.2.2.3)

$$\Delta_l^g S_b = \frac{\Delta_l^g H_b}{T_b} = C_T \qquad (13.1)$$

where C_T is a constant. This is approximately the same for all liquid metallic elements.[1] Figure 13.1 shows a plot of $\Delta_l^g H_b$ vs. T_b, i.e. Trouton's rule, for 62 liquid metallic

[1] In general, liquids which are composed of nonpolar, quasi-spherical molecules or atoms.

The Thermophysical Properties of Metallic Liquids: Volume 2 – Predictive Models. First Edition.
Takamichi Iida and Roderick I. L. Guthrie.
Published in 2015 by Oxford University Press.

elements (data points with solid circles). For the 62 liquid metallic elements determined experimentally, the molar entropy of evaporation has a mean value of 102 J mol^{-1} K^{-1}, namely,

$$\Delta_l^g S_b = \frac{\Delta_l^g H_b}{T_b} = 102\,\mathrm{J\,mol^{-1}\,K^{-1}} \tag{1.10}$$

or

$$\Delta_l^g H_b = 102\,T_b \tag{13.2}$$

Associated data for boiling point temperature and molar evaporation enthalpy of all these metallic elements are given in Tables 17.1 and 17.3, respectively.

Trouton's rule has often been used to predict values of $\Delta_l^g H_b$ for a liquid, knowing the liquid's boiling temperature. Table 13.1 compares experimental values for the evaporation enthalpy of liquid metallic elements with those calculated from Trouton's rule represented by Eq. (13.2). Table 13.1 also lists the δ_i, Δ, and S values needed

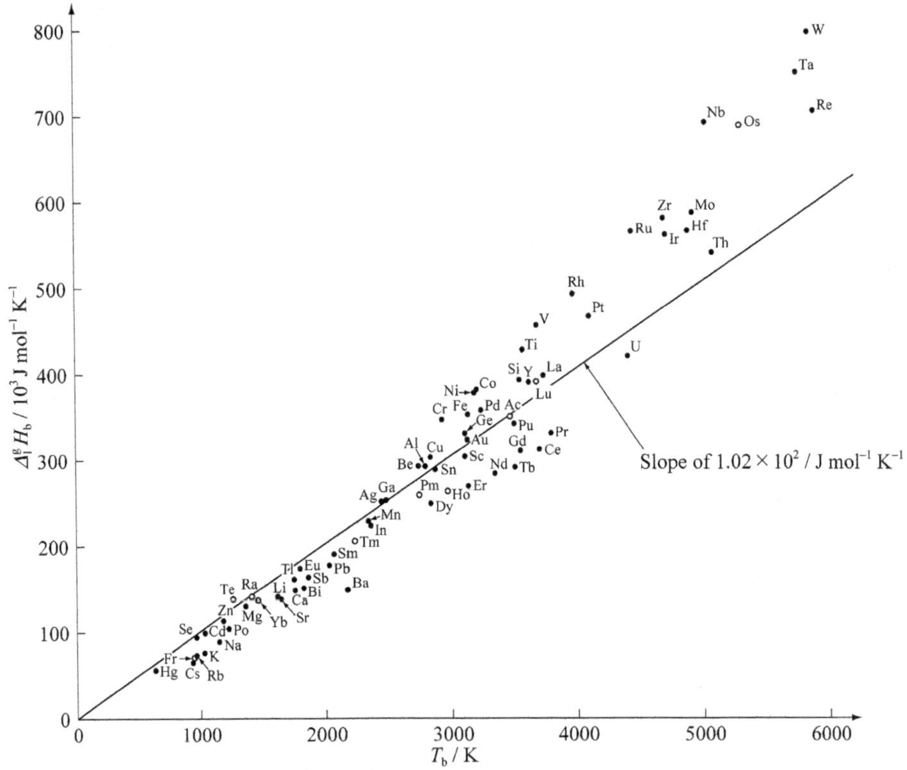

Figure 13.1 *Molar evaporation enthalpy $\Delta_l^g H_b$ plotted against boiling point temperature T_b, i.e. Trouton's rule: ●, experimental; ○, calculated (or predicted).*

for statistical assessment of Trouton's rule. As seen, Trouton's rule performs well with the molar evaporation enthalpy of the 62 liquid metallic elements, giving $\Delta(62)$ and $S(62)$ values of 13.5 per cent and 0.161, respectively. However, as indicated in Figure 13.1, the metals with higher boiling point temperatures are apt to show large deviations from Trouton's rule. As is clear from Table 13.1, the maximum positive deviation from Trouton's rule is 35.9 per cent for niobium, while the minimum negative deviation is 31.7 per cent for barium.

Table 13.1 *Comparison of experimental evaporation enthalpies of liquid metallic elements at their boiling point temperatures with those calculated on the basis of Trouton's rule, together with δ_i, Δ, and S values.*

Element		$\Delta_l^g H_b$ / kJ mol^{-1}		δ_i
		Experimental	Calculated	%
Aluminium	Al	294	285	3.2
Antimony	Sb	165	190	−13.2
Barium	Ba	151	221	−31.7
Beryllium	Be	294	280	5.0
Bismuth	Bi	152	187	−18.7
Cadmium	Cd	100	106	−5.7
Caesium	Cs	65.9	96.3	−31.6
Calcium	Ca	150	179	−16.2
Cerium	Ce	314	378	−16.9
Chromium	Cr	348	300	16.0
Cobalt	Co	383	326	17.5
Copper	Cu	304	289	5.2
Dysprosium	Dy	251	290	−13.4
Erbium	Er	271	320	−15.3
Europium	Eu	176	184	−4.3
Gadolinium	Gd	312	362	−13.8
Gallium	Ga	254	253	0.4
Germanium	Ge	333	317	5.0
Gold	Au	325	319	1.9

continued

Table 13.1 *(continued)*

Element		$\Delta_l^g H_\upsilon$ / kJ mol^{-1}		δ_i
		Experimental	Calculated	%
Hafnium	Hf	571	497	14.9
Indium	In	226	239	−5.4
Iridium	Ir	564	480	17.5
Iron	Fe	354	320	10.6
Lanthanum	La	400	381	5.0
Lead	Pb	180	206	−12.6
Lithium	Li	142	165	−13.9
Magnesium	Mg	132	139	−5.0
Manganese	Mn	230	238	−3.4
Mercury	Hg	57.2	64.2	−10.9
Molybdenum	Mo	590	501	17.8
Neodymium	Nd	284	341	−16.7
Nickel	Ni	380	325	16.9
Niobium	Nb	696	512	35.9
Palladium	Pd	361	330	9.4
Platinum	Pt	469	418	12.2
Plutonium	Pu	344	357	−3.6
Polonium	Po	106	126	−15.9
Potassium	K	77.4	105	−26.3
Praseodymium	Pr	333	387	−14.0
Rhenium	Re	707	599	18.0
Rhodium	Rh	496	405	22.5
Rubidium	Rb	75.2	98.0	−23.3
Ruthenium	Ru	568	451	25.9
Samarium	Sm	192	211	−9.0
Scandium	Sc	305	317	−3.8

Table 13.1 *(continued)*

Element		$\Delta_l^g H_b$ / kJ mol^{-1}		δ_i
		Experimental	Calculated	%
Selenium	Se	94.8	97.7	−3.0
Silicon	Si	395	361	9.4
Silver	Ag	253	248	2.0
Sodium	Na	89.1	118	−24.5
Strontium	Sr	139	169	−17.8
Tantalum	Ta	753	585	28.7
Terbium	Tb	293	357	−17.9
Thallium	Tl	162	178	−9.0
Thorium	Th	544	516	5.4
Tin	Sn	291	293	−0.7
Titanium	Ti	430	363	18.5
Tungsten	W	800	594	34.7
Uranium	U	423	449	−5.8
Vanadium	V	458	375	22.1
Yttrium	Y	393	369	6.5
Zinc	Zn	115	120	−4.2
Zirconium	Zr	582	478	21.8
			Δ(62) %	13.5
			S (62)	0.161

To conclude, Trouton's rule is a useful rough approximation and predicts a rough value for a liquid metallic element's evaporation enthalpy.

13.2.2 Relationship between Evaporation Enthalpy and Cohesive Energy

We showed in Chapter 4 that a good linear relationship exists between the molar evaporation enthalpies $\Delta_l^g H_b$ of liquid metallic elements at their boiling point temperatures and their cohesive energies E_c^0 at 0 K at 1 atm. For 59 metallic elements, the relationship is expressed as (see Subsection 4.1.3.1)

$$\Delta_l^g H_b = 0.889 E_c^0 \tag{4.6}$$

Data for the cohesive energy at 0 K at 1 atm are available for 90 elements, but not for metals such as francium and promethium [1].

Figure 13.2 shows a plot of $\Delta_l^g H_b$ vs. E_c^0 for 59 liquid metallic elements, plus two semimetals (antimony and bismuth), boron, selenium, and sulphur. The relationship for the 64 elements illustrated in Figure 13.2 is described by the following equation:

$$\Delta_l^g H_b = 0.881 E_c^0 \tag{13.3}$$

in which the constant of proportionality was obtained so as to give the minimum S value for the 64 elements.

Table 13.2 gives a comparison of experimental evaporation enthalpies for the 64 liquid elements at their boiling point temperatures with those calculated from Eq. (13.3), together with the corresponding δ_i, Δ, and S values. Equation (13.3) performs well with

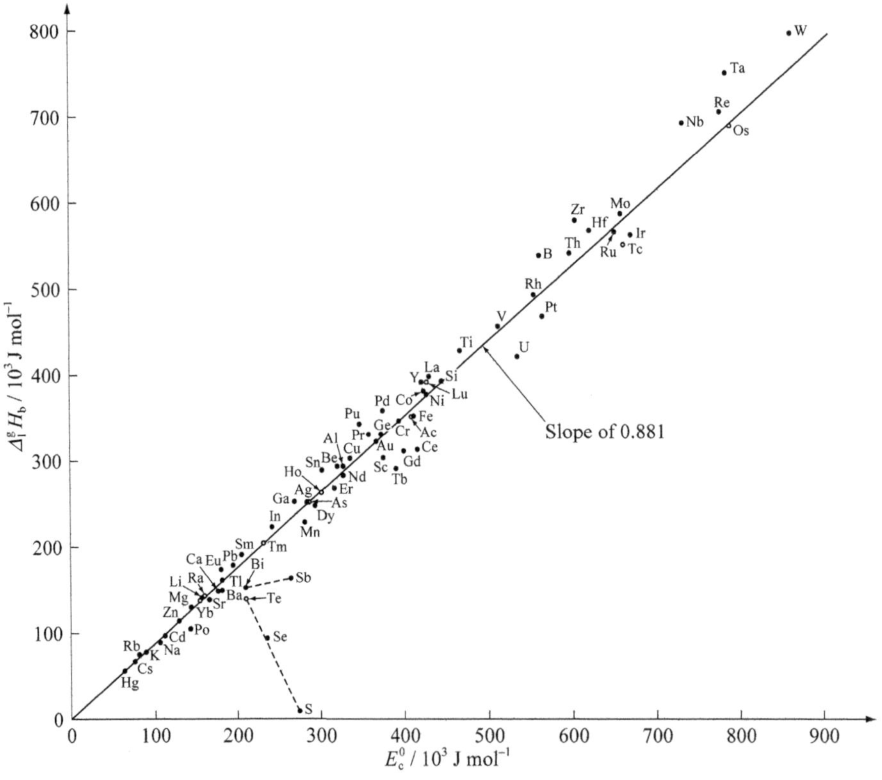

Figure 13.2 *Molar evaporation enthalpy $\Delta_l^g H_b$ plotted against cohesive energy E_c^0: ●, experimental; ○, calculated.*

$\Delta(64)$ and $S(64)$ values of 7.8 per cent and 0.157, respectively. Nevertheless, selenium and sulphur show large negative deviations from the relationship given by Eq. (13.3). For liquid metallic elements, excluding semimetals and the periodic Group VIA elements (sulphur and selenium), the maximum positive deviation from the relationship represented by Eq. (13.3) is 12.4 per cent for plutonium, and the minimum negative deviation is 16.5 per cent for polonium; it would be estimated that even these values fall, or almost fall, within the ranges of their experimental uncertainties.

Table 13.2 *Comparison of experimental evaporation enthalpies of liquid elements at their melting point temperatures with those calculated from Eq. (13.3), together with δ_i, Δ, and S values.*

Element		$\Delta_l^g H_b$ / kJ mol^{-1}		δ_i %
		Experimental	Calculated	
Aluminium	Al	294	288	2.1
Antimony	Sb	165	233	−29.2
Barium	Ba	151	161	−6.2
Beryllium	Be	294	282	4.3
Bismuth	Bi	152	185	−17.8
Boron	B	541	494	9.5
Cadmium	Cd	100	98.7	1.3
Caesium	Cs	65.9	68.4	−3.7
Calcium	Ca	150	157	−4.5
Cerium	Ce	314	367	−14.4
Chromium	Cr	348	348	0
Cobalt	Co	383	374	2.4
Copper	Cu	304	296	2.7
Dysprosium	Dy	251	259	−3.1
Erbium	Er	271	279	−2.9
Europium	Eu	176	158	11.4
Gadolinium	Gd	312	352	−11.4
Gallium	Ga	254	239	6.3
Germanium	Ge	333	328	1.5

continued

Table 13.2 *(continued)*

Element		$\Delta_l^g H_b$ / kJ mol^{-1}		δ_i %
		Experimental	Calculated	
Gold	Au	325	324	0.3
Hafnium	Hf	571	547	4.4
Indium	In	226	214	5.6
Iridium	Ir	564	590	−4.4
Iron	Fe	354	364	−2.7
Lanthanum	La	400	380	5.3
Lead	Pb	180	173	4.0
Lithium	Li	142	139	2.2
Magnesium	Mg	132	128	3.1
Manganese	Mn	230	248	−7.3
Mercury	Hg	57.2	57.3	−0.2
Molybdenum	Mo	590	580	1.7
Neodymium	Nd	284	289	−1.7
Nickel	Ni	380	377	0.8
Niobium	Nb	696	643	8.2
Palladium	Pd	361	331	9.1
Platinum	Pt	469	497	−5.6
Plutonium	Pu	344	306	12.4
Polonium	Po	106	127	−16.5
Potassium	K	77.4	79.4	−2.5
Praseodymium	Pr	333	315	5.7
Rhenium	Re	707	683	3.5
Rhodium	Rh	496	488	1.6
Rubidium	Rb	75.2	72.4	3.9
Ruthenium	Ru	568	573	−0.9
Samarium	Sm	192	181	6.1

Table 13.2 *(continued)*

Element		$\Delta_l^g H_b$ / kJ mol^{-1}		δ_i
		Experimental	Calculated	%
Scandium	Sc	305	331	−7.9
Selenium	Se	94.8	209	−54.6
Silicon	Si	395	393	0.5
Silver	Ag	253	250	1.2
Sodium	Na	89.1	94.3	−5.5
Strontium	Sr	139	146	−4.8
Sulphur	S	9.30	242	−96.2
Tantalum	Ta	753	689	9.3
Terbium	Tb	293	344	−14.8
Thallium	Tl	162	160	1.3
Thorium	Th	544	527	3.2
Tin	Sn	291	267	9.0
Titanium	Ti	430	412	4.4
Tungsten	W	800	757	5.7
Uranium	U	423	472	−10.4
Vanadium	V	458	451	1.6
Yttrium	Y	393	372	5.6
Zinc	Zn	115	115	0
Zirconium	Zr	582	531	9.6
			Δ(64) %	7.8
			S (64)	0.157

13.2.3 Relationship between Evaporation Enthalpy and Melting Point Temperature

In order to accurately predict values of several important thermophysical properties of liquid metallic elements, in the year 2005, the authors [2–4] introduced two common parameters, denoted by ξ_T and ξ_E, which characterize the liquid metallic atoms' state.

These parameters are linked to the Einstein–Lindemann model and the Gitis–Mikhailov model for the velocity of sound U_m in the liquid metallic elements at their melting point temperatures. The models incorporating $\xi_T^{1/2}$ and $\xi_E^{1/2}$ are, respectively, given by

$$U_m = 9.197\left(\frac{\xi_T R T_m}{M}\right)^{1/2} \tag{5.32}$$

$$U_m = \left(\frac{2\xi_E \Delta_l^g H_b}{M}\right)^{1/2} \tag{5.26}$$

Combination of those two equations gives

$$\Delta_l^g H_b = 3.516 \times 10^2 \xi\, T_m, \quad (\text{in J mol}^{-1}) \tag{4.10}$$

where $\xi \equiv \xi_T/\xi_E$. As shown in Figure 4.6, the values of ξ vary periodically with atomic number. Thus, on the basis of Eq. (4.10), we can predict values for evaporation enthalpy $\Delta_l^g H_b$ of liquid metallic elements, using their melting point temperature data.

13.3 Predictions of Evaporation Enthalpies

Using Eqs. (4.6) and (4.10),[2] the authors now predict or calculate values for the molar evaporation enthalpies of 12 liquid metallic elements at their melting point temperatures. Tables 13.3 and 13.4 list calculated values for their molar evaporation enthalpy and the values for the physical quantities or the parameters used for the calculations, respectively. Incidentally, the values of ξ for the elements, given in Table 13.4, were obtained through interpolation using the periodic relationship indicated in Figure 4.6 or Figure 13.3 (data points with solid circles). The values of ξ needed for interpolation are given in Table 13.5; these ξ values were determined from the relationship described by Eq. (4.10), i.e. $\xi = \Delta_l^g H_b / 3.516 \times 10^2\, T_m$, using experimental data on $\Delta_l^g H_b$ and T_m (these data are given in Tables 17.1 and 17.3).

As is obvious from Table 13.3, the difference in the values calculated from Eq. (4.6) or (13.3) and Eq. (4.10) is not more than 15 per cent except for four metallic elements, i.e. radium, tellurium, thulium, and ytterbium. With the exception of these four metallic elements, the mean values of the calculated evaporation enthalpies using Eq. (4.6) or (4.13) and Eq. (4.10) have been adopted (or recommended). For radium, tellurium, thulium, and ytterbium, there exist large differences in the values calculated from Eq. (4.6) or (13.3) and Eq.(4.10). We now consider the following points. Values of the evaporation enthalpies at the boiling point temperatures must be smaller than those of

[2] Using Eq. (13.3) for arsenic and tellurium.

Table 13.3 *Calculated evaporation enthalpies of some liquid metallic elements at their boiling point temperatures.*

Element		$\Delta_l^g H_b$ / kJ mol^{-1}			Diff† %
		Eq. (4.6)	Eq. (4.10)	Adop‡	
Actinium*	Ac	364	341	353	6.5
Arsenic	As	251††	255	253	1.6
Francium*	Fr	–	70.6	70.6	–
Holmium	Ho	268	262	265	2.3
Lutetium	Lu	380	404	392	6.1
Osmium	Os	701	681	691	2.9
Promethium*	Pm	–	261	261	–
Radium*	Ra	142	240	142	51.3
Technetium*	Tc	588	515	552	13.2
Tellurium##	Te	186††	139	139	28.8
Thulium	Tm	207	309	207	39.5
Ytterbium##	Yb	137	207	137	40.7

† Diff.(%) = $|A-B| / \frac{1}{2}(A+B)$, where A and B are values calculated from Eqs. (4.6) (or Eq. (13.3)) and (4.10), respectively.
‡ Adopted value.
* Radioactive.
†† Calculated from Eq. (13.3).
Since $\Delta_g^l H_b$ values are given, on the basis of Eq. (5.26b) (i.e. $\xi_E^{1/2} = (M/2\Delta_l^g H_b)^{1/2} U_m$), the $\xi_E^{1/2}$ value can be easily calculated: $(\xi_E^{1/2})_{Te} = 0.602$, $(\xi_E^{1/2})_{Yb} = 1.012$. (Experimental sound velocity, U_m, values for tellurium and ytterbium are listed in Table 17.7).

the cohesive energy at 0 K at 1 atm and, furthermore, metals will probably not show large deviations from the relationship represented by Eq. (4.6). (Note that radium, thulium, and ytterbium are metals which have body-centred cubic or hexagonal structures in the solid state at their melting point temperatures (see Table 3.2.)). Thus, the authors have adopted the values of 142, 207, and 137 kJ mol^{-1} for the boiling point evaporation enthalpies of radium, thulium, and ytterbium, respectively. For tellurium, the authors have adopted the value of 139 kJ mol^{-1} for its $\Delta_l^g H_b$, so that sulphur, selenium, and tellurium (i.e. the Group VIA elements; atomic number increases by 18 in order) lie on their own straight line, as shown in Figure 13.2. Incidentally, no data for E_c^0 are available for francium and promethium.

Table 13.4 *Values of the physical quantities or the parameters used for predicting evaporation enthalpies $\Delta_l^g H_b$ of some liquid metallic elements.*

Element		E_c^0 kJ mol^{-1}	T_m K	ξ ($\equiv \xi_T/\xi_E$)
Actinium*	Ac	410	1323	0.733
Arsenic	As	285.3	1090	0.665
Francium*	Fr	–	300	0.669
Holmium	Ho	302	1747	0.426
Lutetium	Lu	428	1936	0.593
Osmium	Os	788	3306	0.586
Promethium*	Pm	–	1443	0.515
Radium*	Ra	160	973	0.701
Technetium*	Tc	661	2443	0.600
Tellurium	Te	211	723	0.545
Thulium	Tm	233	1818	0.483
Ytterbium	Yb	154	1092	0.538

* Radioactive.

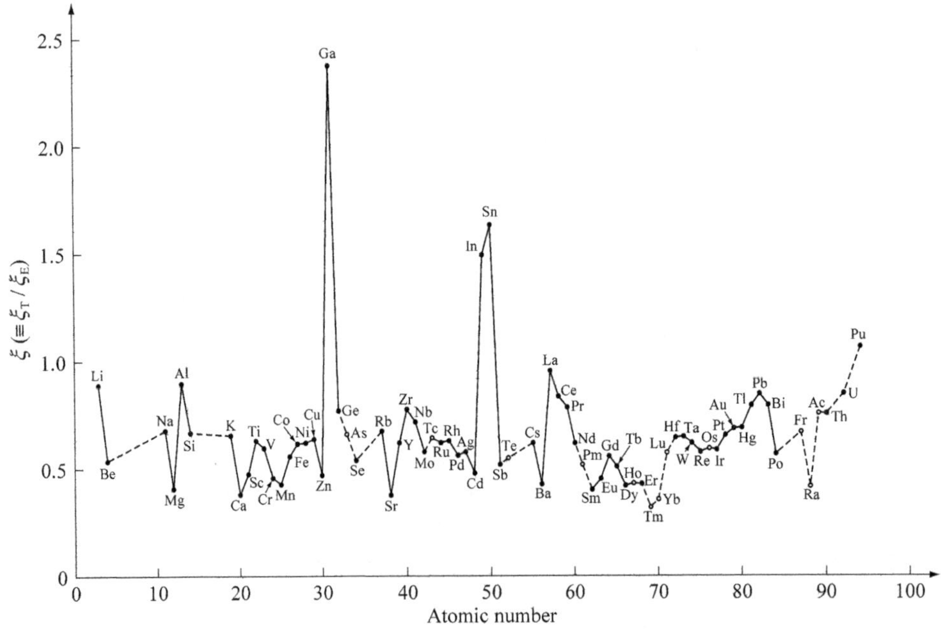

Figure 13.3 *Parameter ξ plotted against atomic number: ●, experimental; ○, calculated.*

Table 13.5 *Values of* $\xi(\equiv \xi_T/\xi_E)$ *for liquid metallic elements at their melting point temperatures.*

Element		ξ
Aluminium	Al	0.896
Antimony	Sb	0.519
Barium	Ba	0.429
Beryllium	Be	0.536
Bismuth	Bi	0.794
Cadmium	Cd	0.479
Caesium	Cs	0.622
Calcium	Ca	0.383
Cerium	Ce	0.834
Chromium	Cr	0.454
Cobalt	Co	0.616
Copper	Cu	0.637
Dysprosium	Dy	0.424
Erbium	Er	0.428
Europium	Eu	0.457
Gadolinium	Gd	0.560
Gallium	Ga	2.385
Germanium	Ge	0.783
Gold	Au	0.691
Hafnium	Hf	0.648
Indium	In	1.496
Iridium	Ir	0.590
Iron	Fe	0.556
Lanthanum	La	0.955
Lead	Pb	0.852
Lithium	Li	0.890

continued

Table 13.5 *(continued)*

Element		ξ
Magnesium	Mg	0.407
Manganese	Mn	0.431
Mercury	Hg	0.694
Molybdenum	Mo	0.579
Neodymium	Nd	0.624
Nickel	Ni	0.625
Niobium	Nb	0.720
Palladium	Pd	0.562
Platinum	Pt	0.653
Plutonium	Pu	1.072
Polonium	Po	0.572
Potassium	K	0.654
Praseodymium	Pr	0.787
Rhenium	Re	0.581
Rhodium	Rh	0.631
Rubidium	Rb	0.684
Ruthenium	Ru	0.620
Samarium	Sm	0.405
Scandium	Sc	0.478
Selenium	Se	0.546
Silicon	Si	0.667
Silver	Ag	0.583
Sodium	Na	0.683
Strontium	Sr	0.376
Tantalum	Ta	0.651
Terbium	Tb	0.512
Thallium	Tl	0.798

Table 13.5 *(continued)*

Element		ξ
Thorium	Th	0.765
Tin	Sn	1.639
Titanium	Ti	0.630
Tungsten	W	0.616
Uranium	U	0.854
Vanadium	V	0.597
Yttrium	Y	0.623
Zinc	Zn	0.472
Zirconium	Zr	0.778

13.4 Relationship between Evaporation Enthalpy of Liquid Metallic Elements and Other Thermophysical Properties or Quantities

Figure 13.1 shows the calculated evaporation enthalpies, listed in Table 13.3, together with those derived experimentally, against their absolute boiling point temperatures, i.e. Trouton's rule, for liquid metallic elements. We see that the ten calculated $\Delta_l^g H_b$ values are suitably located in Trouton's rule. Figure 13.2 shows a plot of $\Delta_l^g H_b$ against E_c^0 for a large number of elements, i.e. the ten calculated and the 64 experimental data points. The graph shows a good linear correlation between the two variables, except for the semimetals (antimony and bismuth) and the periodic Group VIA elements (sulphur, selenium, and tellurium). As mentioned previously, these three elements in the Group VIA lie on their own straight line. Figure 13.3 shows the periodic variation in the parameter $\xi\,(\equiv \xi_T/\xi_E)$ at the melting point temperature with increase in the atomic number of the metallic element; all of the six metals in the Group IIA occupy the valleys of the plot. The Group IA metals have almost the same ξ values, excluding lithium. In Figure 13.4, the evaporation enthalpies of 72 liquid metallic elements, i.e. 62 experimental and ten calculated $\Delta_l^g H_b$ values, are plotted against their liquid ranges $(T_b - T_m)$. As can be seen, this empirical relationship also holds roughly true for the ten calculated values.

Figure 13.5 shows the periodic variation of $(-\langle\Phi\rangle_b/\Delta_l^g H_b)$, $(-\frac{\langle\Phi\rangle}{\Delta_l^g H_b} = 1 - \frac{RT_b}{\Delta_l^g H_b}$; see Subsection 4.1.2), with the atomic number of the metallic element. As can be seen from the figure, the d-block transition metals in the Groups IVB through VIIIB have the highest values, excluding manganese, while the s-block alkali metals, or the Group IA metals, have the lowest values. The periodic variation of $(RT_b/\Delta_l^g H_b)$ plotted against the atomic number is given in Figure 13.6.

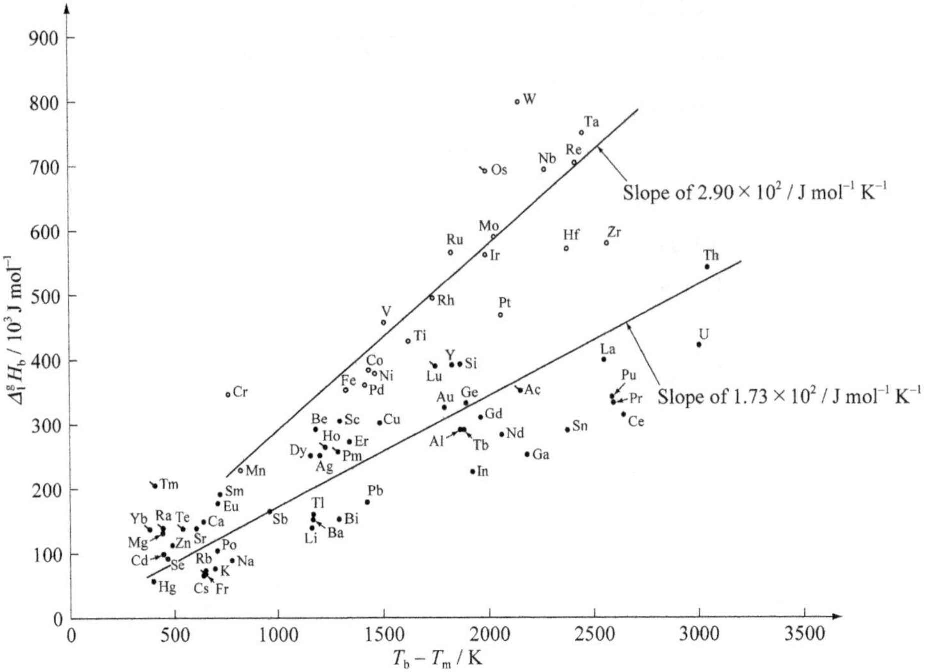

Figure 13.4 *Molar evaporation enthalpy $\Delta_l^g H_b$ plotted against liquid range ($T_b - T_m$): ●, ○, calculated (cf. Figure 1.7).*

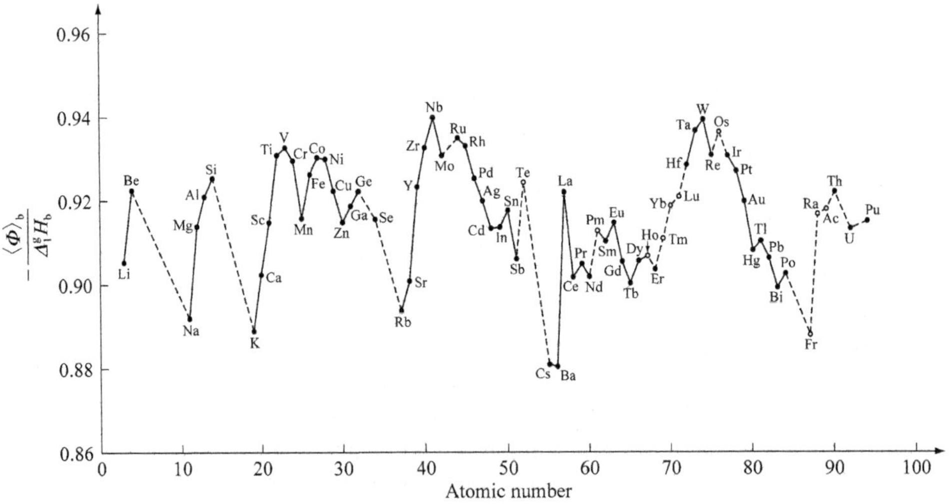

Figure 13.5 *$-\langle \Phi \rangle_b / \Delta_l^g H_b$ plotted against atomic number: ●, experimental; ○, calculated.*

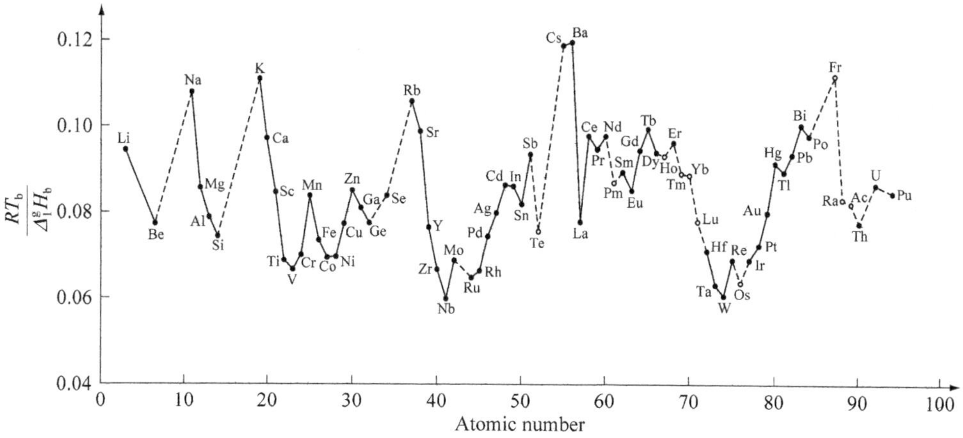

Figure 13.6 *$RT_b / \Delta_l^g H_b$ plotted against atomic number (cf. Eq. (1.11), or Trouton's rule): ●, experimental; ○, calculated.*

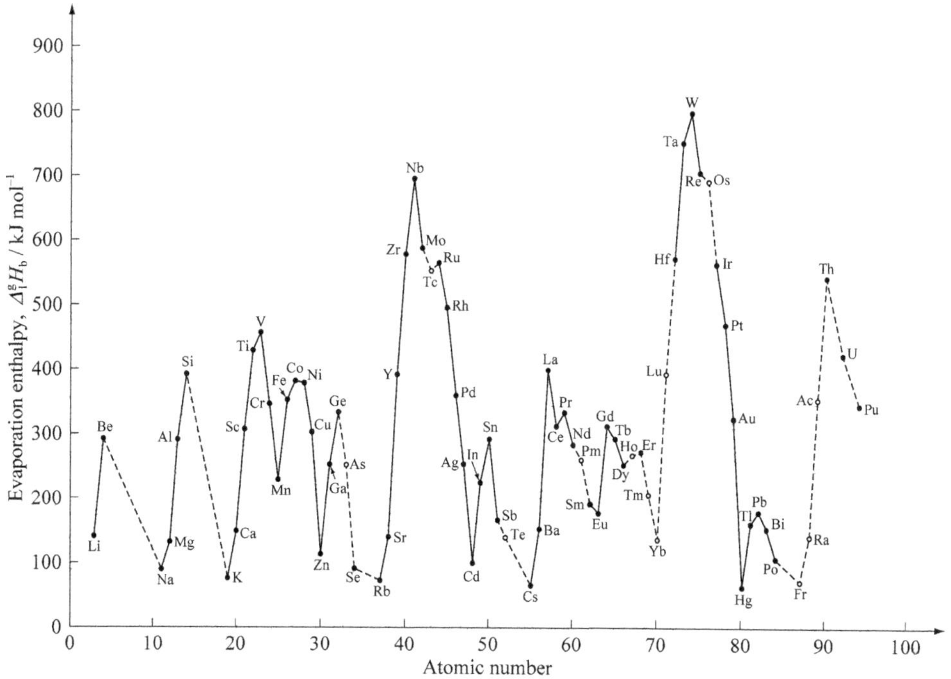

Figure 13.7 *Variation of evaporation enthalpy $\Delta_l^g H_b$ of liquid metallic elements with atomic number: ●, experimental; ○, calculated.*

Values of the boiling point evaporation enthalpy as a function of the atomic number are shown in Figure 13.7, in which the 62 experimental and the 12 calculated $\Delta_l^g H_b$ values for liquid metallic elements are also plotted. There is, clearly, a periodic variation in $\Delta_l^g H_b$ values, the Group IVA metallic elements occupying the peaks and the Groups IA and IIB metals (i.e. alkali metals and zinc group metals) occupying the valleys of the plot. On the whole, the highest values are apparent for the *d*-block transition metals in the Groups IVB through VIIIB, except for manganese. Incidentally, diagonal relationships are noticeable with the following pairs: lithium–magnesium and beryllium–aluminium. These two pairs each have approximately the same values for $\Delta_l^g H_b$.

The uncertainties in the predicted (or calculated) evaporation enthalpy values of the liquid metallic elements at their boiling point temperatures, listed in Table 13.3, can be estimated to be ca. 10 to 20 per cent, and are expected to fall, almost fall, within the ranges of uncertainties of any experimental measurement made in the future.

Generally speaking, accurate and reliable predictions of the thermophysical properties of metallic liquids will be obtained through a combination of plural models.

REFERENCES

1. C. Kittel, *Introduction to Solid State Physics*, 7th ed., John Willey & Sons, Inc., 1996, p.57.
2. T. Iida, R.I.L. Guthrie, and M. Isac, in *ICS Proceedings of the 3rd International Congress on Science and Technology of Steelmaking*, Association for Iron and Steel Technology, Charlotte, NC, 2005, p. 3.
3. T. Iida, R.I.L. Gurhrie, and M. Isac, in *ICS Proceedings of the 3rd International Congress on Science and Technology of Steelmaking*, Association for Iron and Steel Technology, Charlotte, NC, 2005, p. 57.
4. T. Iida, R. Guthrie, M. Isac, and N. Tripathi, *Metall. Mater., Trans. B*, 37 (2006), 403.

14

Predictions of Surface Tension

14.1 Introduction

The surface tension of metallic liquids is one of the most important thermophysical properties. In any and every liquid metallic processing operation, the behaviour of metallic liquids is closely related to their surface tension (e.g. Marangoni convection, Weber number, mass transfer coefficient (cf. Figure 6.1)). As such, accurate and reliable surface tension values are indispensable. Thus, numerous studies on the surface tension of metallic liquids have been made for a long time, from the viewpoint of chemical physics, or materials science, and technology.

We repeat here that, in the field of materials process science, both accuracy and universality are required of any model for calculating or predicting the thermophysical properties of liquid metallic elements. On the basis of such ideas, in this chapter, we discuss the modified Schytil model, and using that model, we then calculate values for the melting point surface tensions of the liquid metallic elements, for those not yet available.

14.2 Calculations or Predictions of Metallic Liquid Surface Tensions

14.2.1 The Modified Schytil Model

There have been a great number of research papers on the surface tension of metallic liquids up until now, because of its importance in the area of materials science and technology. Nevertheless, it is still difficult to identify a truly successful model for which both accuracy and universality apply. Although the Skapski and the Schytil models for the surface tension of liquid metallic elements are virtually universal, the weakness of these models is that agreement with experiment is not necessarily satisfactory from the standpoint of materials process science. In view of this, in Chapter 6, the present authors proposed a modified Skapski model and a modified Schytil model.

The Thermophysical Properties of Metallic Liquids: Volume 2 – Predictive Models. First Edition.
Takamichi Iida and Roderick I. L. Guthrie.
Published in 2015 by Oxford University Press.

The modified Schytil model, expressed in terms of the parameters γ_0 and $\xi_T^{1/2}$, is given by

$$\gamma_m = 3.190\gamma_0 \frac{\xi_T^{1/2} R T_m}{V_m^{2/3}} \qquad (6.39)^1$$

$$\gamma_0 \equiv k_0 C_A$$

where k_0 is a dimensionless numerical factor ($k_0 \approx 0.117$)[2] and C_A is the Andrade coefficient.

The modified Skapski model is expressed as[3]

$$\gamma_m = 7.542 \times 10^{-2} \gamma_0 \frac{\xi_E}{\xi_T^{1/2}} \frac{\Delta_l^g H_b}{V_m^{2/3}} \qquad (6.38)$$

Equations (6.38) and (6.39) provide the same results of calculations for the surface tension of liquid metallic elements.

14.2.2 Predictions of the Melting Point Surface Tensions of Ten Liquid Metallic Elements

A large number of experimentally derived surface tension data have been reported for liquid metallic elements; at present, experimental surface tensions are available for some 67 liquid metallic elements. Moreover, the accuracy of the experimental data is comparatively good, composed with accuracies in other thermophysical properties of liquid metallic elements. Taking everything into consideration, several thermophysical properties of liquid metallic elements can be linked quantitatively to their surface tension. It is also possible to accurately calculate or predict other thermophysical property values, using surface tension data available (e.g. the modified Einstein–Lindemann model, and the modified Fowler–Born–Green relation).

As mentioned previously, a large number of experimental surface tension data are given, but surface tensions have not been measured experimentally for all liquid metallic elements. Thus, we calculate surface tension values for ten liquid metallic elements, i.e. actinium, arsenic, astatine, neptunium, polonium, promethium, protactinium, radium, technetium, and thulium. Unfortunately, there exists little information on their material properties related to the liquid state. With the exception of the melting point temperature and melting point density of arsenic, neither material constants nor physical quantities may be available for these metallic elements.[4] Accordingly, on the basis of Eq. (6.39), we calculate their surface tension values through interpolation, using the periodic variations

[1] $\frac{9.197}{R^{1/2}} k_0 C_A = 3.190\gamma_0$.

[2] See Eq. (6.37).

[3] Combination of Eqs. (4.10) (i.e. $\Delta_l^g H_b = \frac{9.197^2 R}{2} \frac{\xi_T}{\xi_E} T_m$) and (6.39) gives Eq. (6.38).

[4] Needless to say, the molar mass (or the relative atomic mass) and atomic number are known for all elements.

of the parameters appearing in Eq. (6.39) with atomic number.[5] From the standpoint of materials process science, this expedient approach is useful and has already produced fruitful results (see Chapters 13 and 15).

As is obvious from Eq. (6.39), if accurate values of the parameters or physical quantities, $\gamma_0(\equiv k_0 C_A)$, $\xi_T^{1/2}$, T_m and $V_m^{2/3}$, are given, accurate estimates for surface tension values can easily be calculated. Comparatively accurate experimental values of T_m and $V_m(\equiv M/\rho_m)$ are already known for a large number of liquid metallic elements. These are listed in Tables 17.1 and 17.4, respectively. Values of the common parameter $\xi_T^{1/2}$ for 41 liquid metallic elements, determined using experimental sound velocities, and those 26 liquid metals, determined using predicted sound velocities, are given in Tables 5.5 and 11.4, respectively. Values of γ_0 for 41 liquid metallic elements, which are calculated on experimental data for $\xi_T^{1/2}$, T_m, and $V_m^{2/3}$ (using Eq. (6.39) or (6.39′)), are illustrated in Figure 6.7, and these are given in Table 14.1.

Table 14.1 *Values of γ_0 ($\equiv k_0 C_A$) appearing in Eqs. (6.39) and (6.38).*

Element		γ_0 10^{-8} kg$^{1/2}$ m s^{-1} K$^{-1/2}$ mol$^{-1/6}$
Aluminium	Al	2.252
Antimony	Sb	1.307
Barium	Ba	2.094
Beryllium	Be	0.984
Bismuth	Bi	1.652
Cadmium	Cd	2.024
Caesium	Cs	1.923
Calcium	Ca	1.736
Cerium	Ce	2.911
Chromium	Cr	*1.559*†
Cobalt	Co	1.783
Copper	Cu	1.646
Dysprosium	Dy	*1.708*

continued

[5] The periodic variations of the thermophysical properties or physical quantities of elements with atomic number suggest that discussions are needed, based on a quantum mechanical (electron) theory. However, it is extremely difficult to present a calculable theoretical formula that gives sufficient accuracy, combined with a universality based on a quantum mechanical theory to this subject.

Table 14.1 *(continued)*

Element		γ_0 10^{-8} $kg^{1/2}$ m s^{-1} $K^{-1/2}$ $mol^{-1/6}$
Erbium	Er	*1.625*
Europium	Eu	*1.559*
Francium	Fr	*2.128*
Gadolinium	Gd	*1.807*
Gallium	Ga	2.777
Germanium	Ge	1.342
Gold	Au	1.394
Hafnium	Hf	*1.443*
Holmium	Ho	*1.674*
Indium	In	2.209
Iridium	Ir	*1.378*
Iron	Fe	1.772
Lanthanum	La	2.287
Lead	Pb	1.625
Lithium	Li	2.784
Lutetium	Lu	*1.617*
Magnesium	Mg	1.845
Manganese	Mn	1.677
Mercury	Hg	2.826
Molybdenum	Mo	1.313
Neodymium	Nd	*2.080*
Nickel	Ni	1.677
Niobium	Nb	*1.364*
Osmium	Os	*1.284*
Palladium	Pd	*1.388*
Platinum	Pt	1.359
Plutonium	Pu	1.846

Table 14.1 *(continued)*

Element		γ_0 10^{-8} kg$^{1/2}$ m s^{-1} K$^{-1/2}$ mol$^{-1/6}$
Potassium	K	2.114
Praseodymium	Pr	2.196
Rhenium	Re	*1.329*
Rhodium	Rh	*1.407*
Rubidium	Rb	2.069
Ruthenium	Ru	*1.392*
Samarium	Sm	*1.580*
Scandium	Sc	*1.892*
Selenium	Se	1.093
Silicon	Si	1.567
Silver	Ag	1.474
Sodium	Na	2.271
Strontium	Sr	1.801
Tantalum	Ta	1.340
Tellurium	Te	2.200
Terbium	Tb	*1.773*
Thallium	Tl	1.760
Thorium	Th	*1.409*
Tin	Sn	1.924
Titanium	Ti	*2.078*
Tungsten	W	1.288
Uranium	U	*1.792*
Vanadium	V	*1.834*
Ytterbium	Yb	1.595
Yttrium	Y	*1.983*
Zinc	Zn	1.903
Zirconium	Zr	*1.507*

† Numerical figures in italics denote that $\xi_T^{1/2}$ values have been calculated on the basis of the predicted sound velocity values (see Table 11.4).

We now add 26 more γ_0 values for liquid metals. Equation (6.39) can be rewritten as

$$\gamma_0 = 3.771 \times 10^{-2} \frac{\gamma_m V_m^{2/3}}{\xi_T^{1/2} T_m} \qquad (6.39')$$

Using $\xi_T^{1/2}$ values determined on the basis of predicted sound velocities (see Table 11.4), values of γ_0 for 26 liquid metals have been calculated from Eq. (6.39′). For γ_m, $V_m^{2/3}$, and T_m, experimentally obtained values were used for the calculations. Incidentally, Table 14.2 lists values of $V_m^{2/3}$ for 70 liquid elements, and Figure 14.1 shows the periodic variation in their $V_m^{2/3}$ values plotted against atomic number. Calculated values of γ_0 (on the basis of predicted sound velocities) are also given in Table 14.1.

Table 14.2 *Values of $V_m^{2/3}$ for liquid elements at their melting point temperatures.*

Element		$V_m^{2/3}$ 10^{-4} m^2 mol$^{-2/3}$
Aluminium	Al	5.039
Antimony	Sb	7.065
Arsenic	As	5.919
Barium	Ba	11.96
Beryllium	Be	3.052
Bismuth	Bi	7.561
Boron	B	3.001
Cadmium	Cd	5.809
Caesium	Cs	17.34
Calcium	Ca	9.517
Cerium	Ce	7.602
Chromium	Cr	4.090
Cobalt	Co	3.862
Copper	Cu	3.981
Dysprosium	Dy	7.220
Erbium	Er	7.095
Europium	Eu	9.569

Table 14.2 *(continued)*

Element		$V_m^{2/3}$ 10^{-4} m^2 mol$^{-2/3}$
Francium	Fr	20.80
Gadolinium	Gd	7.417
Gallium	Ga	5.065
Germanium	Ge	5.585
Gold	Au	5.050
Hafnium	Hf	6.376
Holmium	Ho	7.170
Indium	In	6.429
Iridium	Ir	4.520
Iron	Fe	3.980
Lanthanum	La	8.165
Lead	Pb	7.221
Lithium	Li	5.642
Lutetium	Lu	6.843
Magnesium	Mg	6.160
Manganese	Mn	4.498
Mercury	Hg	5.987
Molybdenum	Mo	4.734
Neodymium	Nd	7.749
Nickel	Ni	3.806
Niobium	Nb	5.212
Osmium	Os	4.473
Palladium	Pd	4.685
Platinum	Pt	4.740
Plutonium	Pu	5.990
Potassium	K	13.08

continued

Table 14.2 *(continued)*

Element		$V_m^{2/3}$ 10^{-4} m^2 $mol^{-2/3}$
Praseodymium	Pr	7.686
Rhenium	Re	4.611
Rhodium	Rh	4.495
Rubidium	Rb	14.93
Ruthenium	Ru	4.413
Samarium	Sm	7.442
Scandium	Sc	6.560
Selenium	Se	7.294
Silicon	Si	4.976
Silver	Ag	5.124
Sodium	Na	8.504
Strontium	Sr	11.10
Sulphur	S	6.774
Tantalum	Ta	5.271
Tellurium	Te	7.851
Terbium	Tb	7.294
Thallium	Tl	6.871
Thorium	Th	7.875
Tin	Sn	6.611
Titanium	Ti	5.124
Tungsten	W	4.765
Uranium	U	5.748
Vanadium	V	4.486
Ytterbium	Yb	8.731
Yttrium	Y	7.684
Zinc	Zn	4.625
Zirconium	Zr	6.350

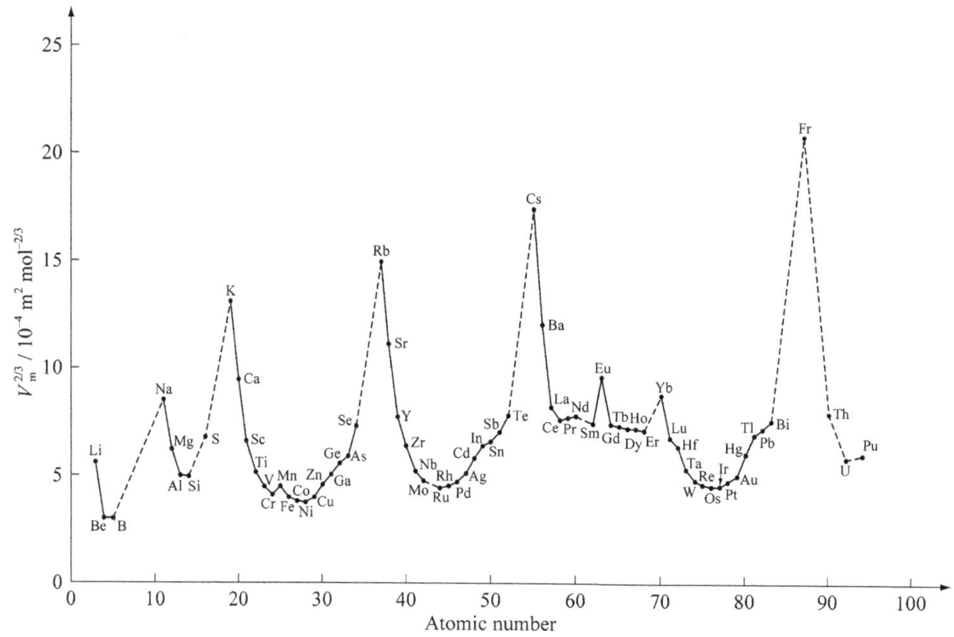

Figure 14.1 *Plot of* $V_m^{2/3}$ *against atomic number for a large number of liquid metallic elements.*

Figure 14.2 provides a plot of γ_0 against atomic number for 67 liquid metallic elements listed in Table 14.1. As can be seen, calculated γ_0 values appear to be suitably located in Figure 14.2; in view of γ_0 values, the *d*-block transition metals can be classified into two groups: the *d*-block transition metals in period 4 (group 1)[6] and the *d*-block transition metals in periods 5 and 6 (group 2).[7] The metals in each group have approximately the same γ_0 values; in other words, each group has its own γ_0 value. It can be estimated that this result stems from the influence of the Andrade coefficient C_A values.

Figures 14.3 and 14.4 show comparisons of the melting point surface tensions of 26 liquid metals (plutonium and uranium are included in group 1, and thorium in group 2) based on the plots of $RT_m / V_m^{2/3}$ (i.e. the Schytil model) and $\gamma_0 \xi_T^{1/2} RT_m / V_m^{2/3}$ (i.e. the modified Schytil model represented by Eq. (6.39)). In Figure 14.3, the slope of 4.788×10^{-8} mol$^{1/3}$ was obtained so as to give the minimum S value for the 26 liquid metals plotted in the figure. As such, the Schytil model for the 26 liquid metals is expressed as

$$\gamma_m = 4.778 \times 10^{-8} \frac{RT_m}{V_m^{2/3}} \qquad (14.1)$$

The averaged means of γ_0 values in the groups 1 and 2 are 1.766×10^{-8} and 1.379×10^{-8} kg$^{1/2}$ m s^{-1} K$^{-1/2}$ mol$^{-1/6}$, respectively; these γ_0 values were used for each

[6] Do not confuse with groups in the periodic table. The vertical columns of the periodic table are called groups (or families).

[7] See the periodic table of the (chemical) elements (e.g. Table 10.4).

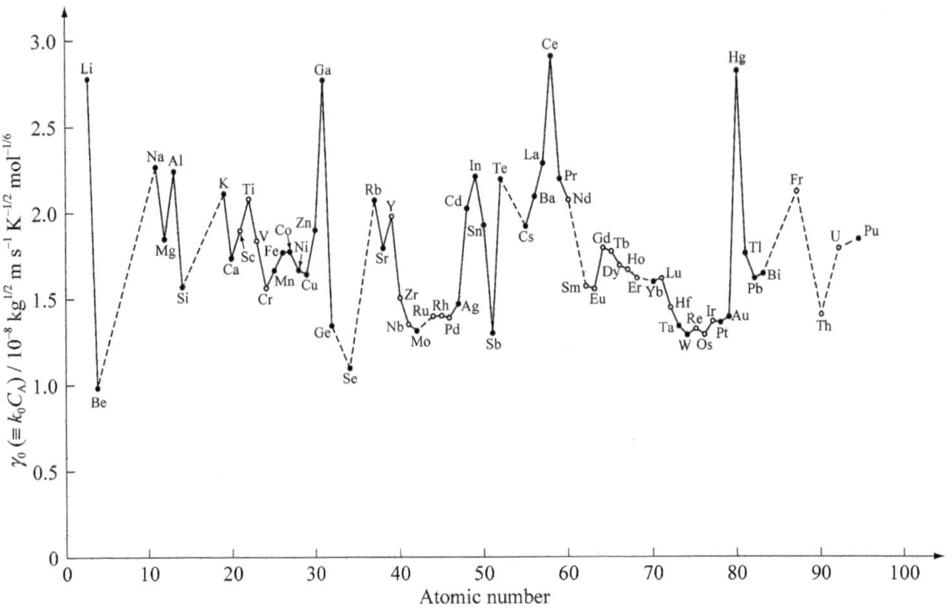

Figure 14.2 *Plot of $\gamma_0(\equiv k_0 C_A)$ against atomic number: ●, calculated using $\xi_T^{1/2}$ values determined on the basis of experimental sound velocities; ○, calculated using $\xi_T^{1/2}$ values determined on the basis of predicted sound velocities.*

individual group in Figure 14.4. As can be seen, while both plots show linear relationships, the latter gives a much better correlation.

Table 14.3 compares experimental surface tension values, taken from Table 17.8, for the 26 liquid metals with those calculated from Eqs. (14.1) and (6.39). As is clear from the table, the modified Schytil model, i.e. Eq. (6.39), performs extremely well, with $\Delta(26)$ and $S(26)$ values of 4.3 per cent and 0.058, respectively; most of the calculated values fall within the ranges of uncertainties associated with experimental measurements. Incidentally, titanium has a large δ_{Ti} value of 17.7 per cent because of the large γ_0 value (probably, its large C_A value) in the group 1 metals, as shown in Table 14.1.

Now, values of the parameters or physical quantities used for this purpose are all available. On the basis of Eq. (6.39), we calculate values for the melting point surface tension γ_{m} for as yet unmeasured liquid metallic elements, mentioned previously, using the method of interpolation. Calculated values of γ_{m} are listed in Table 14.4, in which values of the parameters or physical quantities used for the calculations are also listed. Incidentally, as shown in Table 14.4, values of $\xi_{\mathrm{T}}^{1/2}$ for the ten liquid metallic elements are given, so that we can immediately calculate sound velocities U_{m}, on the basis of Eq. (5.32) i.e. $U_{\mathrm{m}} = 9.197\xi_{\mathrm{T}}^{1/2}(RT_{\mathrm{m}}/M)^{1/2}$; calculated sound velocities are listed in Table 14.5.

Figure 14.5 gives a plot of γ_{m} against $RT_{\mathrm{m}}/V_{\mathrm{m}}^{2/3}$, i.e. the Schytil model, for a large number of liquid metallic elements (i.e. 67 experimental and ten calculated values

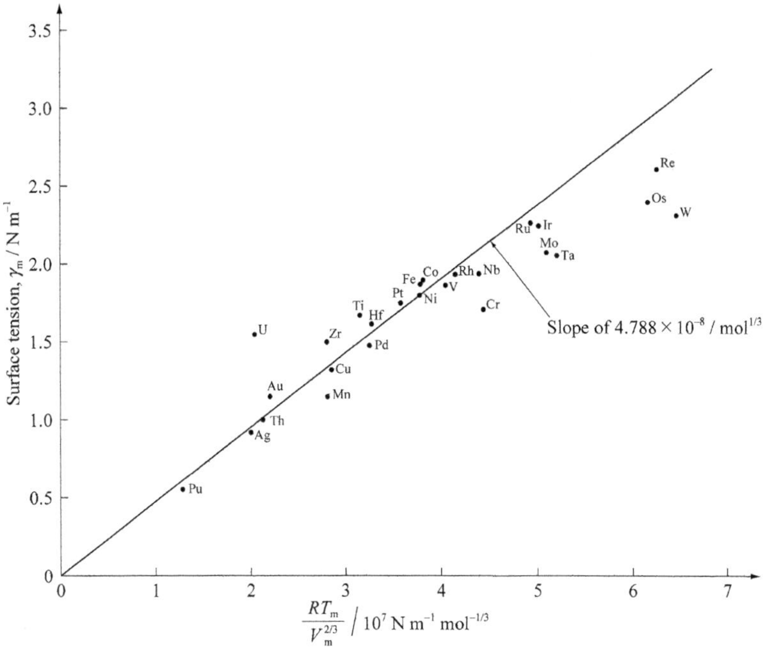

Figure 14.3 *Melting point surface tension of liquid metals plotted against* $RT_m / V_m^{2/3}$*, i.e. the Schytil model.*

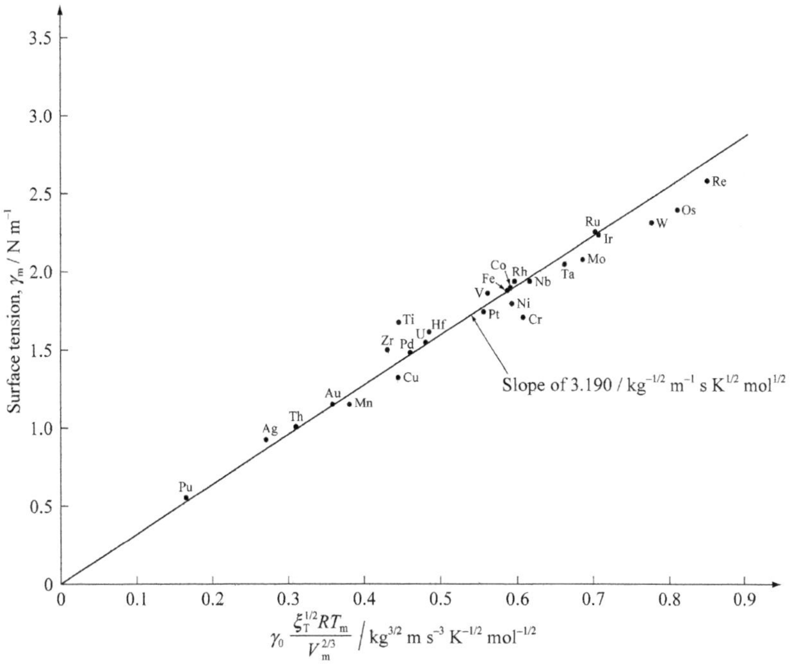

Figure 14.4 *Melting point surface tension of liquid metals plotted against* $\gamma_0 \xi_T^{1/2} RT_m / V_m^{2/3}$*, i.e. the modified Schytil model.*

Table 14.3 *Comparison of experimental and calculated values, on the basis of Eqs. (14.1) and (6.39), for the melting point surface tension of the d-block transition metals in period 4 plus plutonium and uranium (group 1)†, and the d-block transition metals in periods 5 and 6, plus thorium (group 2)‡, in the liquid state, together with δi, Δ, and S values.*

Group	Metal		$(\gamma_m)_{exp}$ N m^{-1}	$(\gamma_m)_{cal}$ / N m^{-1}		δi / %	
				Eq. (14.1)	Eq. (6.39)	Eq. (14.1)	Eq. (6.39)
	Chromium	Cr	1.710	2.122	1.937	−19.4	−11.7
	Cobalt	Co	1.900	1.822	1.882	4.3	1.0
	Copper	Cu	1.320	1.358	1.417	−2.8	−6.8
	Iron	Fe	1.880	1.811	1.873	3.8	0.4
1	Manganese	Mn	1.152	1.344	1.213	−14.3	−5.0
	Nickel	Ni	1.795	1.807	1.891	−0.7	−5.1
	Titanium	Ti	1.670	1.508	1.419	10.7	17.7
	Vanadium	V	1.860	1.937	1.792	−4.0	3.8
	Plutonium	Pu	0.550	0.607	0.526	−9.4	4.6
	Uranium	U	1.552	0.975	1.530	59.2	1.4
	Gold	Au	1.150	1.054	1.138	9.1	1.1
	Hafnium	Hf	1.614	1.565	1.543	3.1	4.6
	Iridium	Ir	2.250	2.394	2.253	−6.0	−0.1
	Molybdenum	Mo	2.080	2.435	2.186	−14.6	−4.8
	Niobium	Nb	1.937	2.100	1.959	−7.8	−1.1
	Osmium	Os	2.400	2.942	2.579	−18.4	−6.9
2	Palladium	Pd	1.482	1.553	1.473	−4.6	0.6
	Platinum	Pt	1.746	1.715	1.772	1.8	−1.5
	Rhenium	Re	2.610	2.986	2.708	−12.6	−3.6
	Rhodium	Rh	1.940	1.981	1.902	−2.1	2.0
	Ruthenium	Ru	2.260	2.352	2.238	−3.9	1.0
	Silver	Ag	0.926	0.960	0.867	−3.5	6.8
	Tantalum	Ta	2.050	2.484	2.109	−17.5	−2.8
	Tungsten	W	2.310	3.087	2.473	−25.2	−6.6
	Zirconium	Zr	1.500	1.334	1.373	12.4	9.2
	Thorium	Th	1.006	1.023	0.985	−1.7	2.1
					Δ(26) %	10.1	4.3
					S (26)	0.157	0.058

† γ_0 value of 1.766×10^{-8} kg$^{1/2}$ m s^{-1} K$^{-1/2}$ mol$^{-1/6}$ was used for the metals in group 1.
‡ γ_0 value of 1.379×10^{-8} kg$^{1/2}$ m s^{-1} K$^{-1/2}$ mol$^{-1/6}$ was used for the metals in group 2.

Table 14.4 *Calculated values for the surface tension of ten liquid metallic elements at their melting point temperatures, together with the physical quantities or the parameters used for calculating their surface tension values.*

Metallic element		$(\gamma_m)_{cal}$ $N\,m^{-1}$	γ_0 $10^{-8}\,kg^{1/2}\,m\,s^{-1}\,K^{-1/2}\,mol^{-1/6}$	$\xi_T^{1/2}$	T_m K	$V_m^{2/3}$ $10^{-4}\,m^2\,mol^{-2/3}$
Actinium*	Ac	0.427	1.649	0.953	1323	12.91
Arsenic	As	0.390	1.218	0.656	1090	5.92†
Astatine*	At	0.190	1.890	0.987	575	14.96
Neptunium*	Np	0.768	1.819	1.035	903	5.869
Polonium*	Po	0.235	1.771	1.099	527	11.56
Promethium*	Pm	0.606	1.830	0.657	1443	7.596
Protactinium*	Pa	1.559	1.601	1.191	2113	6.853
Radium*	Ra	0.244	1.888	0.857	973	17.08
Technetium*	Tc	1.930	1.353	1.007	2443	4.576
Thulium	Tm	0.581	1.610	0.594	1818	7.946

* Radioactive.
† Calculated using an experimental density value.

Table 14.5 *Calculated, or predicted, sound velocities in ten liquid metallic elements at their melting point temperatures, together with the molar mass used for the calculations.*

Metallic element		$(U_m)_{cal}$ $m\,s^{-1}$	M $kg\,mol^{-1}$
Actinium*	Ac	1929	0.22703
Arsenic	As	2098	0.07492
Astatine*	At	1370	0.20999
Neptunium*	Np	1694	0.23705
Polonium*	Po	1464	0.20898
Promethium*	Pm	1727	0.14692
Protactinium*	Pa	3020	0.23104
Radium*	Ra	1491	0.22603
Technetium*	Tc	4240	0.09691
Thulium	Tm	1634	0.16893

* Radioactive.

of γ_m). The slope of 5.023×10^{-8} mol$^{-1/3}$ represents the mean value obtained from that of the respective metallic elements whose surface tensions were determined experimentally. The overall scatter, shown in Figure 14.5, is rather high, which results from the product of the parameters $\gamma_0 (\equiv k_0 C_A)$ and $\xi_T^{1/2}$.

Figure 14.6 shows the periodic variation in values of the melting point surface tension of liquid metallic elements plotted against atomic number. All the metals in Group IA occupy the valleys of the plot. Evidently, the highest values are for the *d*-block transition metals, excluding manganese and copper group transition metals. The *d*-block transition metals all have two outer *s*-electrons and have *d*-electrons in their penultimate shell (see Tables 1.5 and 1.7(b)). In the transition metals, there is additional binding (covalent-type bonds) from the inner electron shells, which is characterized by high binding energy. As we move down a group in the periodic table, the melting point surface tension of liquid metallic elements in Groups IA through IVA, IIB, and IIIB decrease. To give examples, in Group IA metals, Li(0.399) → Na(0.197) → K(0.110) → Rb(0.0896) → Cs(0.069) → Fr(0.062); in the case of Group IIA and IIIB metals, Be(1.100) → Mg(0.577) → Ca(0.363) → Sr(0.296) → Ba(0.272) →

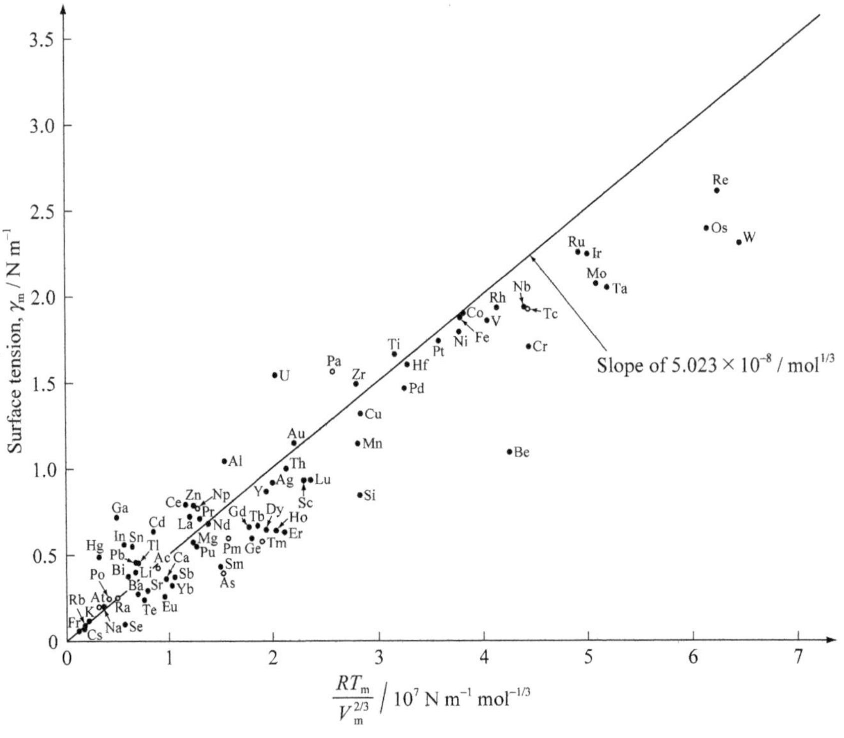

Figure 14.5 *Melting point surface tension plotted against $RT_m / V_m^{2/3}$, i.e. the Schytil model, for a great number of liquid metallic elements: ●, experimental; ○, calculated or predicted.*

Ra(*0.244*) and Sc(0.939) → Y(0.872) → La(0.728) →Ac(*0.427*), where numerical figures in parentheses denote surface tension values in N m^{-1}. These results suggest the calculated γ_m values, denoted by italics, may be reasonable. As can be seen from Figure 14.6, the peak of the *d*-block transition metals in period 5 splits into two, owing to the sinking of technetium; if this is true, the method of interpolation based on the periodic variation in values of γ_m should give only rough predictions of γ_m. Incidentally, values for the melting point surface tensions of liquid metallic elements can be used to calculate, or predict, the temperature coefficients, $d\gamma/dT$, of their respective surface tensions.

Equation (6.39) can be written as follows:

$$\gamma_m = 3.190 k_s \frac{RT_m}{V_m^{2/3}} \tag{14.2a}$$

$$k_s \equiv \gamma_0 \xi_T^{1/2}$$

$$k_s = 3.771 \times 10^{-2} \frac{\gamma_m V_m^{2/3}}{T_m} \tag{14.2b}$$

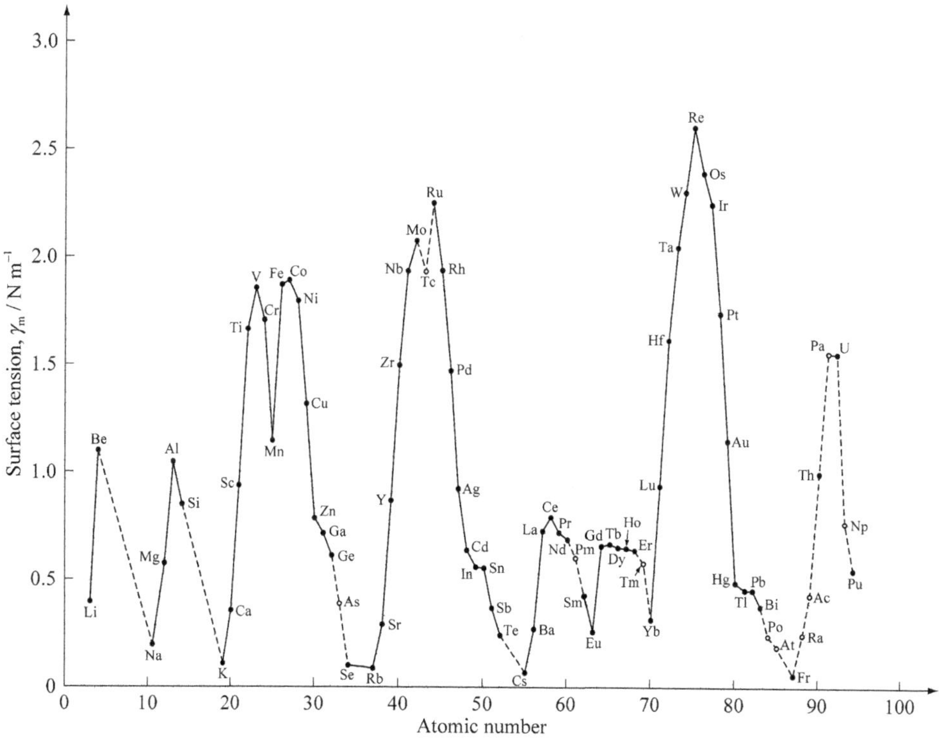

Figure 14.6 *Periodic variation of the melting point surface tension of liquid metallic elements with atomic number: ●, experimental; ○, calculated or predicted (cf. also literature results [1–3]).*

Using Eq. (14.2b), we calculate values of k_s for 77 liquid metallic elements. Calculated values of k_s are given in Table 14.6. Figure 14.7 shows a plot of k_s against atomic number; in the figure, k_s values for the ten liquid metallic elements, listed in Table 14.6 are also plotted. It can be seen that values of k_s also vary periodically with atomic number. Moreover, making a comparison between Figures 11.5 and 14.7, we see that both plots (of the parameters $\xi_T^{1/2}$ and k_s) show similar periodic variations with increase in the atomic number of the elements: aluminium, gallium, indium (Group IIIA metals), mercury, and uranium occupy the major peaks, whereas potassium, rubidium, caesium (Group IA metals), titanium, zirconium, and hafnium (Group IVB metals) the minor peaks. Similarly, silicon, selenium, and tellurium (semiconductors or metalloids)[8] occupy the lowest valleys, whereas calcium, strontium, barium (Group IIA metals), molybdenum, and tungsten (Group VIB metals)[9] the low valleys of the plot. Note that if k_s values of radium and actinium are lowed by ca. 10 per cent and 7 per cent, respectively, i.e. if the calculated values of γ_m for radium and actinium are actually 0.220 and 0.400 in N m^{-1} (the points marked ⌕), respectively (these values are within the ranges of probable uncertainties in the calculations), in Figure 14.7, then all the Group IIA metals occupy the valleys of the plot.

Table 14.6 *Values of* k_s ($\equiv \gamma_0 \xi_T^{1/2} \equiv k_0 C_A \xi_T^{1/2}$) *appearing in Eq. (14.2).*

Element		k_s 10^{-8} kg$^{1/2}$ m s^{-1} k$^{-1/2}$ mol$^{-1/6}$
Actinium	Ac	*1.571*†
Aluminium	Al	2.138
Antimony	Sb	1.094
Arsenic	As	*0.7988*
Astatine	At	*1.864*
Barium	Ba	1.231
Beryllium	Be	0.8116
Bismuth	Bi	2.000
Cadmium	Cd	2.349
Caesium	Cs	1.496
Calcium	Ca	1.168
Cerium	Ce	2.126

[8] Liquid sulphur is a non-metal.
[9] Chromium may be included.

Table 14.6 *(continued)*

Element		k_s 10^{-8} kg$^{1/2}$ m s^{-1} k$^{-1/2}$ mol$^{-1/6}$
Chromium	Cr	1.210
Cobalt	Co	1.565
Copper	Cu	1.460
Dysprosium	Dy	1.047
Erbium	Er	0.9459
Europium	Eu	0.8699
Francium	Fr	1.621
Gadolinium	Gd	1.171
Gallium	Ga	4.566
Germanium	Ge	1.057
Gold	Au	1.638
Hafnium	Hf	1.549
Holmium	Ho	1.006
Indium	In	3.160
Iridium	Ir	1.411
Iron	Fe	1.558
Lanthanum	La	1.882
Lead	Pb	2.072
Lithium	Li	1.871
Lutetium	Lu	1.253
Magnesium	Mg	1.452
Manganese	Mn	1.287
Mercury	Hg	4.712
Molybdenum	Mo	1.282
Neodymium	Nd	1.551
Neptunium	Np	*1.882*

continued

Table 14.6 *(continued)*

Element		k_s 10^{-8} kg$^{1/2}$ m s^{-1} k$^{-1/2}$ mol$^{-1/6}$
Nickel	Ni	1.491
Niobium	Nb	1.384
Osmium	Os	1.224
Palladium	Pd	1.432
Platinum	Pt	1.529
Plutonium	Pu	1.361
Polonium	Po	*1.944*
Potassium	K	1.613
Praseodymium	Pr	1.723
Promethium	Pm	*1.203*
Protactinium	Pa	*1.907*
Radium	Ra	*1.616*
Rhenium	Re	1.312
Rhodium	Rh	1.470
Rubidium	Rb	1.615
Ruthenium	Ru	1.443
Samarium	Sm	0.8957
Scandium	Sc	1.281
Selenium	Se	0.5738
Silicon	Si	0.9469
Silver	Ag	1.449
Sodium	Na	1.703
Strontium	Sr	1.180
Tantalum	Ta	1.239
Technetium	Tc	*1.363*
Tellurium	Te	0.9787

Table 14.6 *(continued)*

Element		k_s $10^{-8}\ kg^{1/2}\ m\ s^{-1}\ k^{-1/2}\ mol^{-1/6}$
Terbium	Tb	1.130
Thallium	Tl	2.061
Thorium	Th	1.477
Thulium	Tm	*0.9562*
Tin	Sn	2.739
Titanium	Ti	1.662
Tungsten	W	1.123
Uranium	U	2.389
Vanadium	V	1.441
Ytterbium	Yb	0.9648
Yttrium	Y	1.407
Zinc	Zn	1.987
Zirconium	Zr	1.688

† Numerical figures in italics denote values calculated on the basis of predicted surface tensions.

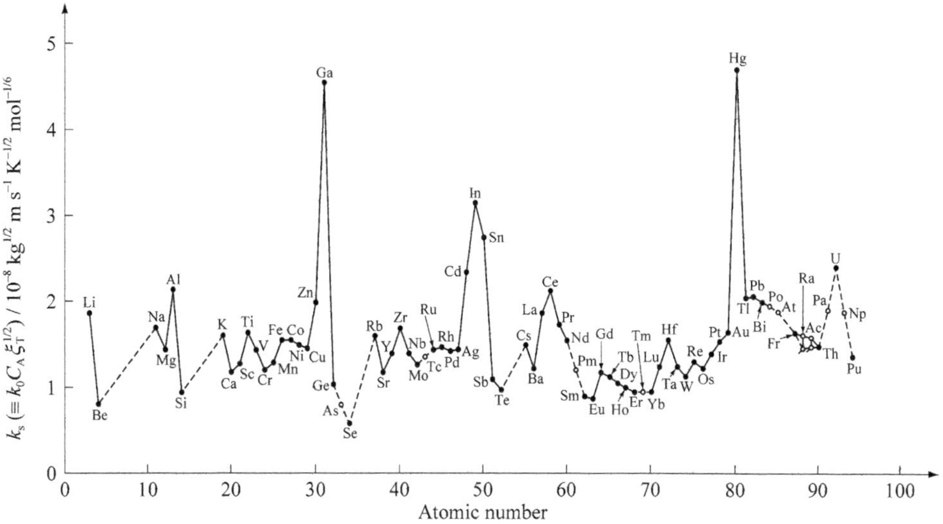

Figure 14.7 *Plot of* $k_s(\equiv k^0 C_A \xi_T^{1/2})$ *vs. atomic number:* ●, *determined from experimental surface tensions;* ○, *determined from calculated or predicted surface tensions (the points marked* ρ *indicate low values ca. by 10 or 7 per cent, compared with those marked* ○ *for radium and actinium).*

Lanthanoid metals (excluding cerium) also show similar periodic variations in values of the two parameters, $\xi_T^{1/2}$ and k_s, with atomic number. Thus, the modified Schytil model, expressed in terms of $\xi_T^{1/2}$ (only one simple correction factor), makes a noticeable improvement over the Schytil model (see Table 6.4). Nevertheless, the following show different variations between the two plots: (i) lithium → beryllium → sodium → magnesium, (ii) lanthanum → cerium → praseodymium, and (iii) bismuth → francium → thorium. In addition, all the *d*-block transition metals have roughly the same values of k_s, but the *d*-block transition metals in period 4 have somewhat different $\xi_T^{1/2}$ values from those in periods 5 and 6. These discrepancies in both plots may be due to values of the Andrade coefficient.

In general, Eq. (6.38) or (6.39) must be used to calculate accurate values for the melting point surface tension of liquid metallic elements.

14.3 Temperature Coefficient of Surface Tension

The temperature coefficient, or the temperature dependence, $d\gamma / dT$, of the surface tension of a liquid is a crucial factor to evaluate the magnitude of Marangoni flows [4] on the surface of a liquid.[10] As such, there have been many studies on the surface tensions of liquids, and their temperature coefficients. However, large discrepancies exist among the experimental data on $d\gamma / dT$ for metallic liquids, even for the common low melting point metallic elements, because of the difficulty in obtaining accurate values for their $d\gamma / dT$'s. Consequently, it is difficult to identify the true capability of models for the temperature coefficient of the surface tension of liquid metallic elements.

Recently, Ishikawa and Paradis [5] checked the validity of the model represented by Eq. (6.42) in Chapter 6 (cf. Subsection 6.3.2 and Sections 6.5 and 6.6) on the basis of experimental data on $d\gamma / dT$ for various liquid metallic elements. They have rewritten Eq. (6.42) in the following from:

$$\left.\frac{d\gamma}{dT}\right|_{T=T_m} = \frac{2}{3}\frac{\Lambda\gamma_m}{\rho_m}\frac{2\underline{\alpha}+1}{\underline{\alpha}} = K\gamma_m\alpha_m \tag{14.3}$$

$$K \equiv -\frac{2(2\underline{\alpha}+1)}{3\underline{\alpha}}$$

where Λ is the temperature dependence $d\rho/dT$ of liquid density ρ, α is the volume expansivity ($\alpha_m \equiv -\Lambda / \rho_m$), $\underline{\alpha}$ is a parameter related to the distance over which the interatomic force extends; $\underline{\alpha}$ is a constant which is roughly the same for all liquid metals ($\underline{\alpha} \sim 0.45$ to 0.65), and subscript m stands for the melting point.

The derivation of Eq. (14.3) is as follows. The surface tension and its temperature dependence for a liquid metallic element at any temperature about its melting point temperature can be expressed by (see Subsection 6.3.2 and Section 6.5), respectively,

[10] If there is a temperature or concentration gradient along the surface of a liquid, Marangoni flow (or convection) will occur.

$$\gamma = \frac{(\pi\beta c)^2 T_m}{2N_A M^{2/3}}\left(\frac{\rho}{\rho_m}\right)^{2/3}\left\{\left(1+\underline{\alpha}\right)\rho^{1/3} - \rho_m^{1/3}\right\}^2 \tag{6.33}$$

and

$$\frac{d\gamma}{dT} = \frac{(\pi\beta c)^2 T_m \Lambda}{3N_A M^{2/3}}\left\{2(1+\underline{\alpha})^2\rho^{1/3}\rho_m^{-2/3} + \rho^{-1/3} - 3(1+\underline{\alpha})\rho_m^{-1/3}\right\} \tag{6.42}$$

At the melting point, these equations can be written as

$$\gamma_m = \frac{(\pi\beta c)^2 T_m \underline{\alpha}^2 \rho_m^{2/3}}{2N_A M^{2/3}} \tag{6.33'}$$

and

$$\left.\frac{d\gamma}{dT}\right|_{T=T_m} = \frac{(\pi\beta c)^2 T_m \Lambda}{3N_A M^{2/3}} \cdot \frac{\underline{\alpha}(2\underline{\alpha}+1)}{\rho_m^{1/3}} \tag{6.42'}$$

Combination of these two equations lead to

$$\left.\frac{d\gamma}{dT}\right|_{T=T_m} = K\gamma_m\alpha_m, \quad (\alpha_m \equiv -\Lambda/\rho_m) \tag{14.3}$$

$$K \equiv -\frac{2(2\underline{\alpha}+1)}{3\underline{\alpha}}$$

Ishikawa and Paradis concluded on the basis of their experimental results and (literature) data on alkali metals (given that a plot of $d\gamma/dT$ against $\gamma_m\ \alpha_m$, their measured data show the same tendency as literature data on alkali metals which satisfy Eq. (14.3)), that the temperature dependence of the surface tension of liquid metals can be estimated (on the basis of Eq. (14.3)) if γ_m and α_m are known.

Incidentally, an approach based on the principle of corresponding states also gives a very similar result to the above relation, i.e. Eq. (14.3) (see Section 6.5).

We can obtain another form of expression for $d\gamma/dT$. Substituting Eq. (12.1) into Eq. (14.3), we have

$$\left.\frac{d\gamma}{dT}\right|_{T=T_m} = 1.182\times10^{-2} K \left.\frac{\gamma_{G,T} C_P \gamma}{RT\xi_T}\right|_{T=T_m} \tag{14.4}$$

Similarly, substitution of Eq. (12.2) into Eq. (14.3) leads to

$$\left.\frac{d\gamma}{dT}\right|_{T=T_m} = K \left.\frac{\gamma_{G,E} C_P \gamma}{2\Delta_l^g H_b \xi_E}\right|_{T=T_m} \tag{14.5}$$

In conclusion, the models represented by Eqs. (14.3) through (14.5), or Eq. (6.42), can be basically used to calculate values for the temperature coefficient of the surface tension of liquid metallic elements (cf. Table 6.5).

REFERENCES

1. K. Ogino, *The Chemistry of Surfaces and Interfaces at High Temperatures* (*Kōon Kaimen Kogaku*), Volume 1, Agne Gijutsu Sentā, Tokyo, 2008, p.133.
2. B.J. Keene, *Int. Mater. Rev.*, 38 (1993), 157.
3. B.C. Allen, in *Liquid Metals: Chemistry and Physics*, edited by S.Z. Beer, Marcel Dekker, New York, 1972, p. 161.
4. K. Mukai, in *Fundamentals of Metallurgy*, edited by S. Seetharaman, Woodhead Publishing, Cambridge, 2005, p. 237.
5. T. Ishikawa and P.-F. Paradis, in *High-Temperature Measurements of Materials*, H. Fukuyama and Y. Waseda (eds.), Springer, Berlin, 2009, p. 173.

15

Predictions of Viscosity

15.1 Introduction

All operations in liquid metallic processing, such as refining, casting, and crystal growth, involve fluids in motion. In other words, this is based on the utilization of a fluid's characteristic properties (e.g. its lack of rigidity, its capacity to flow, and a high atomic mobility). As a measure of resistance to flow that a liquid offers, or as a measure of the friction among atoms, molecular viscosity is, therefore, an essential physical property in discussing the flow behaviour of the fluid or the dynamic behaviour of atoms. This chapter deals with the molecular viscosity of metallic liquids.[1]

From the viewpoint of materials process science, in which both accuracy and universality are required of any model for the thermophysical properties of the metallic liquids, we discuss three models: the Andrade-type, the Fowler–Born–Green (FBG), and the modified Fowler–Born–Green (MFBG) models. Both the Andrade-type and the MFBG models perform very well for the melting point viscosities of liquid metallic elements. On the basis of the two models, viscosities for 22 liquid metals (e.g. technetium, gadolinium, polonium, radium, neptunium) are calculated or predicted. Calculated viscosities are expected to fall, or almost fall, within the range of uncertainties associated with experimental measurements.

15.2 Viscosity Models for Accurate Predictions

15.2.1 The Andrade-Type Model

The Andrade-type model, which gives the melting point viscosity μ_m of monatomic liquids, is expressed as

$$\mu_m = C_A \frac{(MT_m)^{1/2}}{V_m^{2/3}} \tag{7.14}$$

[1] The present book of two volumes is devoted to the 'molecular viscosity' of metallic liquids, and does not consider 'turbulent viscosities', which are flow dependent, and significantly much larger.

The Thermophysical Properties of Metallic Liquids: Volume 2 – Predictive Models. First Edition.
Takamichi Iida and Roderick I. L. Guthrie.
Published in 2015 by Oxford University Press.

where C_A is called the Andrade coefficient, M is the molar mass, and V_m is the molar volume at the melting point temperature T_m. The Andrade coefficient C_A is an empirical constant which is roughly equal to 1.80×10^{-7} kg$^{1/2}$m s^{-1} K$^{-1/2}$ mol$^{-1/6}$ for pure liquid metals. Thus, Eq.(7.14) for pure liquid metals can be written as follows:

$$\mu_m = 1.80 \times 10^{-7} \frac{(MT_m)^{1/2}}{V_m^{2/3}} \tag{7.15}$$

This equation is called the Andrade formula (or the Andrade model). As already mentioned in Chapter 7, the Andrade formula performs well with $\Delta(48)$ and $S(48)$ values of 14.9 per cent and 0.182, respectively.

Strictly speaking, however, values of the Andrade coefficient C_A vary periodically with atomic number Z, the Groups IA and IIB metals occupying the peaks and the semiconductors (germanium and silicon in Group IVA) and semimetals (antimony and bismuth in Group VA) the valleys of the C_A vs. Z plot. Using this periodic relationship, Iida et al. [1] determined C_A values of several liquid metals such as molybdenum and yttrium, and further, on the basis of the Andrade-type model represented by Eq.(7.14), predicted their viscosities. After they reported the predicted viscosity values, experimental values for molybdenum [2] and yttrium [3] became available. Table 15.1 shows a comparison between experimental and predicted values. As can be seen, Eq.(7.14), i.e. the Andrade-type model, provides much better result as compared with those of the Andrade model; the predicted values of μ_m for molybdenum and yttrium fall within the range of uncertainties associated with experimental measurements, which are estimated to be ca. 15 per cent.

If an accurate value of the Andrade coefficient is given, an accurate melting point viscosity can easily be calculated.

15.2.2 The Modified Fowler–Born–Green Relation

We recall the viscosity models described in Chapter 7: the FBG and the MFBG relations.

At the melting point temperature, the FBG relation, based on statistical mechanics, can be expressed in the form

Table 15.1 *Comparison between experimental and predicted values for the melting point viscosity of liquid molybdenum and yttrium, together with δ_i and C_A values.*

Metal		$(\mu_m)_{exp}$ mPa s	$(\mu_m)_{cal}$ / mPa s		δ_i / %		C_A 10^{-7} kg$^{1/2}$ m s^{-1} mol$^{-1/6}$
			Eq. (7.14)†	Eq. (7.15)‡	Eq. (7.14)†	Eq. (7.15)‡	
Molybdenum	Mo	5.6	5.39	6.34	3.9	−12	1.53
Yttrium	Y	3.6	3.42	2.96	5.3	22	2.08

† The Andrade-type model.
‡ The Andrade formula.

$$\mu_{\rm m} = \frac{16}{15}\left(\frac{m}{kT_{\rm m}}\right)^{1/2}\gamma_{\rm m} \tag{7.3}$$

or

$$\mu_{\rm m} = 0.370\left(\frac{M}{T_{\rm m}}\right)^{1/2}\gamma_{\rm m} \tag{7.4}$$

where m is the atomic mass, k is the Boltzmann constant, γ is the surface tension, M is the molar mass (the atomic mass multiplied by Avogadro constant $N_{\rm A}$ is equal to the molar mass, i.e. $mN_{\rm A} = M$), and subscript m stands for the melting point. The FBG relation performs reasonably with liquid metallic elements, giving $\Delta(37)$ and $S(37)$ values of 32.7 per cent and 0.400, respectively (see Table 7.7); however, these Δ and S values are generally unsatisfactory from the standpoint of materials process science. Since the surface tension is a thermodynamic property, whereas the viscosity is a transport property, a simple connection between the two properties will probably give only an approximate relation. Consequently, some physical quantity linking the two properties into a closer relationship with each other may be needed. Thus, Iida et al. [1] proposed a mathematical relation between the surface tension and the viscosity through the velocity of sound, as mentioned in Chapter 7.

The following is a brief outline on the derivation of the mathematical relation. At the melting point temperature,

$$\gamma_{\rm m} = k_0 \mu_{\rm m} U_{\rm m} \tag{6.37}$$

or

$$\mu_{\rm m} = \frac{\gamma_{\rm m}}{k_0 U_{\rm m}} \tag{15.1}$$

where k_0 is a dimensionless numerical factor, U is the velocity of sound, and subscript m denotes the melting point. Substitution of Eq.(5.32)[2] into Eq.(15.1) leads to

$$\begin{aligned}\mu_{\rm m} &= \frac{1}{9.197R^{1/2}k_0}\left(\frac{M}{\xi_{\rm T} T_{\rm m}}\right)^{1/2}\gamma_{\rm m} \\ &= C_0\left(\frac{M}{\xi_{\rm T} T_{\rm m}}\right)^{1/2}\gamma_{\rm m}\end{aligned} \tag{7.41}$$

$$C_0 = \frac{1}{9.197R^{1/2}k_0} = 0.03771k_0^{-1}\ \ ({\rm kg}^{-1/2}\,{\rm m}^{-1}{\rm s}\,{\rm K}^{1/2}\,{\rm mol}^{1/2})$$

This equation can be called the MFBG relation.

As is evident from Eq.(15.1) or (7.41), if accurate values of k_0, or C_0, $\gamma_{\rm m}$, and $U_{\rm m}$, or $\xi_{\rm T}^{1/2}$, can be deduced, accurate viscosities $\mu_{\rm m}$ are readily calculable.

[2] Equation (5.32) is expressed as

$$U_{\rm m} = 9.197\left(\frac{\xi_{\rm T} R T_{\rm m}}{M}\right)^{1/2}$$

where $\xi_{\rm T}$ is the common parameter, M is the molar mass, and R is molar gas constant ($R = 8.314\,{\rm J\,mol^{-1}\,K^{-1}}$).

15.3 Performance of the Modified Fowler–Born–Green Relation

On the basis of the MFBG relation, in order to accurately predict values of the viscosity of liquid metallic elements, then accurate values of their respective numerical factors k_0 are needed. Unfortunately, uncertainties in experimentally derived viscosity data are estimated to be ca. 1 to 20 per cent, so that it would be reasonable to refrain from detailed discussions. The data points with closed circles plotted in Figure 15.1,[3] on the whole, suggest that values of k_0 may be roughly divided into several groups: (1) the *s*-block

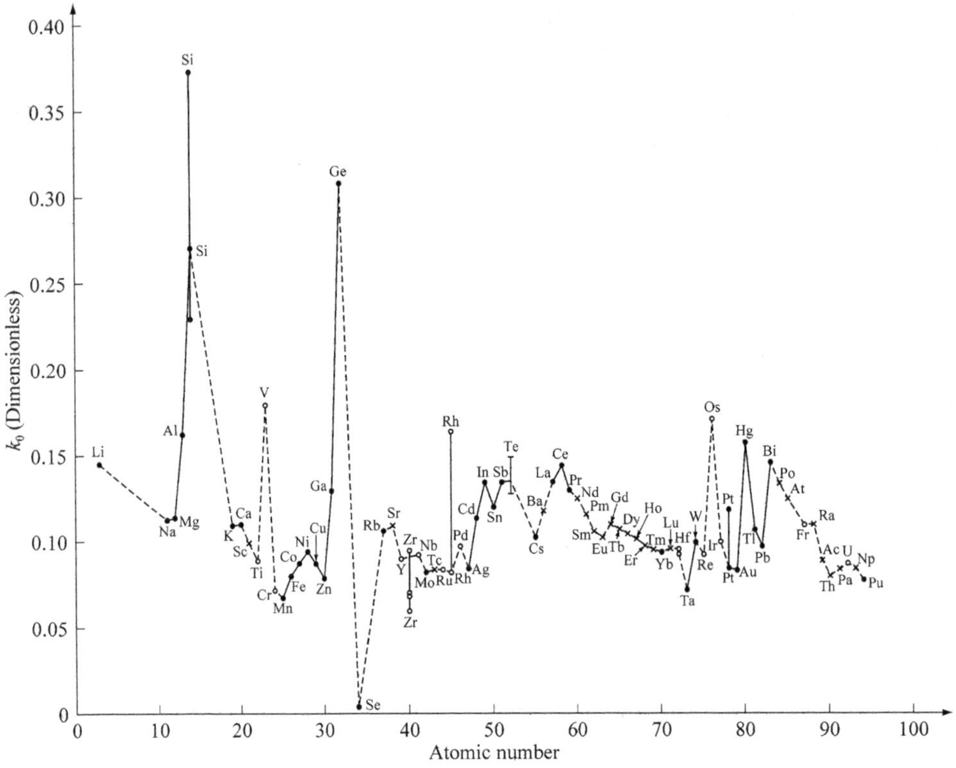

Figure 15.1 *Plot of the numerical factor k_0 appearing in Eq. (6.37) (i.e. $\gamma_m = k_0\, \mu_m U_m$, or $k_0 = \gamma_m / \mu_m U_m$) against atomic number:* ●, I, *calculated using experimental data;* ○, *calculated using predicted sound velocities;* ×, *calculated using both predicted viscosities and sound velocities, except for barium and strontium; calculated using predicted viscosities and experimental sound velocities only for barium and strontium. Points linked by a vertical line indicate values calculated from two, three, or four different experimental viscosities for a single metallic element.*

[3] The data points with closed circles indicate that values of k_0 were determined using experimentally derived data for all the physical quantities, or parameters, i.e. the melting point viscosity μ_m, surface tension γ_m, and sound velocity U_m, appearing in Eq. (6.37).

metals (excluding lithium), (2) the *d*-block transition metals, (3) the *p*-block metallic elements (excluding thallium, lead, and anomalous metallic elements), (4) rare earth metals, and (5) anomalous metallic elements (e.g. germanium, silicon, selenium, etc.).

The following are the values of k_0 ($C_0 \equiv 0.03771\, k_0^{-1}/\text{kg}^{-1/2}\text{m}^{-1}\,\text{s}\,\text{K}^{1/2}\text{mol}^{1/2}$; see Eq. (7.41)) for the *s*-block metals and the *d*-block transition metals, i.e. the above-mentioned groupings 1 and 2.

The *s*-block metals (grouping 1)			The *d*-block transition metals (grouping 2)				
	k_0			k_0		k_0	
Ca	0.1108		Co	0.0873	Ag	0.0855	
Cs	0.1023		Cu	0.0876	Ta	0.0722	
Mg	0.1136		Au	0.0834	W	0.1006	
K	0.1092		Fe	0.0799	(Mn 0.0681)††		
Rb	0.1063		Mo	0.0825			
Na	0.1122		Ni	0.0944			
Ba	0.1189†		Pt	0.0849			
Mean value	0.1105	($C_0 = 0.341$‡)			Mean value	0.0858	($C_0 = 0.439$‡)

† Interporated value between caesium and lanthanum.
‡ The units of $\text{kg}^{-1/2}\,\text{m}^{-1}\,\text{s}\,\text{K}^{1/2}\,\text{mol}^{1/2}$.
†† Excluded, because of an anomalous metal in the *d*-block transition metals.

Using the mean values for k_0 and C_0, we have for the melting point viscosities μ_m for the *s*-block metals:

$$\mu_m = \frac{\gamma_m}{0.1105\, U_m} \tag{15.2}$$

or

$$\mu_m = 0.341\left(\frac{M}{\xi_T T_m}\right)^{1/2} \gamma_m \tag{15.3}$$

For the *d*-block transition metals:

$$\mu_m = \frac{\gamma_m}{0.0858 U_m} \tag{15.4}$$

or

$$\mu_m = 0.439\left(\frac{M}{\xi_T T_m}\right)^{1/2} \gamma_m \tag{15.5}$$

Incidentally, the values of k_0 for a large number of liquid metallic elements are listed in Table 15.2.

Table 15.2 *Values of the numerical factors C_A, k_0, C_0, and k_v for various liquid metallic elements.*

Element		$(\mu_m)_{exp}$ 10^{-3} Pa s	C_A 10^{-7} kg$^{1/2}$ m s^{-1} K$^{-1/2}$ mol$^{-1/6}$	k_0	C_0 kg$^{-1/2}$ m^{-1} s K$^{1/2}$ mol$^{1/2}$	k_v
Aluminium	Al	1.38	1.39	0.1626	0.232	0.660
Antimony	Sb	1.44	0.970	0.1349	0.280	0.903
Bismuth	Bi	1.59	1.13	0.1465	0.257	0.573
Cadmium	Cd	2.50	1.78	0.1139	0.331	0.770
Caesium	Cs	0.686	1.88	0.1023	0.369	1.278
Calcium	Ca	1.10	1.57	0.1108	0.340	1.365
Cerium	Ce	3.25	2.02	0.1443	0.261	0.966
Chromium	Cr	5.7	2.19	*0.0712*†	*0.530*	1.843
Cobalt	Co	5.4	2.04	0.0873	0.432	1.329
Copper	Cu	4.38	1.88	0.0876	0.430	1.310
Francium	Fr	0.765	1.95	*0.1094*	*0.345*	1.223
Gallium	Ga	1.95	2.15	0.1292	0.292	0.479
Germanium	Ge	0.73	0.44	0.3088	0.122	0.419
Gold	Au	5.37	1.67	0.0834	0.452	1.039
Hafnium	Hf	5.2	1.57	*0.0921*	0.409	1.031
		5.0	1.51	*0.0956*	0.394	0.991
Indium	In	1.79	1.64	0.1348	0.280	0.528

Iridium	Ir	7.0	1.38	*0.0995*	*0.379*	0.999
Iron	Fe	5.6	2.22	0.0799	0.472	1.449
Lanthanum	La	2.66	1.69	0.1354	0.279	0.914
Lead	Pb	2.59	1.68	0.0969	0.389	0.823
Lithium	Li	0.602	1.91	0.1455	0.259	1.042
Magnesium	Mg	1.25	1.63	0.1136	0.332	1.140
Manganese	Mn	5	2.46	0.0681	0.554	1.949
Mercury	Hg	2.048	1.79	0.1580	0.239	0.387
Molybdenum	Mo	5.6	1.59	0.0825	0.457	1.263
Nickel	Ni	4.7	1.78	0.0944	0.399	1.213
Niobium	Nb	4.5	1.47	*0.0930*	*0.405*	1.079
Osmium	Os	4.2	0.75	*0.1713*	*0.220*	0.623
Palladium	Pd	4.22	1.42	*0.0979*	*0.385*	1.008
Platinum	Pt	6.74	1.60	0.0849	0.444	1.066
		4.82	1.14	0.1187	0.318	0.763
Plutonium	Pu	5.95	2.39	0.0774	0.487	1.788
Potassium	K	0.537	1.94	0.1092	0.345	1.223
Praseodymium	Pr	2.85	1.68	0.1305	0.289	0.993
Rhenium	Re	7.9	1.44	*0.0926*	*0.407*	1.114

continued

Table 15.2 *(continued)*

Element		$(\mu_m)_{exp}$ 10^{-3} Pa s	C_A 10^{-7} kg$^{1/2}$ m s^{-1} K$^{-1/2}$ mol$^{-1/6}$	k_0	C_0 kg$^{-1/2}$ m^{-1} s K$^{1/2}$ mol$^{1/2}$	k_v
Rhodium	Rh	5	1.48	*0.0949*	*0.397*	1.026
		2.9	0.86	*0.1637*	*0.230*	0.595
Rubidium	Rb	0.674	1.95	0.1063	0.355	1.229
Ruthenium	Ru	6.1	1.66	*0.0840*	*0.449*	1.170
Selenium	Se	24.8	29.0	0.00378	9.98	51.4
Silicon	Si	0.8	0.58	0.2710	0.139	0.622
		0.58	0.42	0.3739	0.101	0.451
		0.94	0.68	0.2307	0.163	0.731
Silver	Ag	3.88	1.72	0.0855	0.441	1.210
Sodium	Na	0.695	2.03	0.1122	0.336	1.210
Tantalum	Ta	8.6	1.86	0.0722	0.522	1.527
Tellurium	Te	1.8–2.1	1.5–1.7	0.1494–0.1280	0.253–0.295	1.531–1.786
Thallium	Tl	2.59	1.64	0.1074	0.351	0.809
Tin	Sn	1.87	1.60	0.1205	0.313	0.593
Titanium	Ti	4.42‡	2.35	*0.0885*	*0.426*	1.439
Tungsten	W	7.0	1.28	0.1006	0.375	1.160
Uranium	U	6.57	2.06	*0.0869*	*0.434*	0.879

Vanadium	V	2.4	1.02	*0.1796*	*0.210*	0.721
Ytterbium	Yb	2.67	1.70	0.0941	0.401	1.789
Yttrium	Y	3.6	2.19	*0.0905*	0.417	1.584
Zinc	Zn	3.50	2.41	0.0791	0.477	1.232
Zirconium	Zr	4.7	2.14	*0.0703*	*0.536*	1.292
		4.83	2.20	*0.0685*	*0.551*	1.328
		3.5	1.59	*0.0945*	*0.399*	0.962
		5.45	2.48	*0.0607*	*0.621*	1.498

† Numeric figures in italics denote values calculated using predicted sound velocities.
‡ From Paradis et al. [4].

Table 15.3 *Comparison of experimental melting point viscosities of the d-block liquid transition metals plus plutonium and uranium (i.e. actinoid metals) with those calculated on the basis of the Andrade, the Fowler–Born–Green, and the modified Fowler–Born–Green models, together with* δ_i, Δ, *and S values.*

Metal		$(\mu_m)_{exp}$ mPa s	$(\mu_m)_{cal}$ / mPa s			δ_i / %		
			Andrade	FBG	MFBG	Andrade	FBG	MFBG
Chromium	Cr	5.7	4.69	3.09	4.72	22	84	21
Cobalt	Co	5.4	4.76	4.06	5.48	13	33	−1.5
Copper	Cu	4.38	4.20	3.34	4.47	4.3	31	−2.0
Gold	Au	5.37	5.79	5.17	5.22	−7.3	3.9	2.9
Hafnium	Hf	5.1†	5.97	5.04	5.57	−15	1.2	−8.4
Iridium	Ir	7.0	9.10	7.00	8.11	−23	0	−14
Iron	Fe	5.6	4.55	3.86	5.21	23	45	7.5
Manganese	Mn	5	3.66	2.56	3.97	37	95	26
Molybdenum	Mo	5.6	6.34	4.43	5.38	−12	26	4.1
Nickel	Ni	4.7	4.76	3.87	5.17	−1.3	21	−9.1
Niobium	Nb	4.5	5.52	4.17	4.87	−18	7.9	−7.6
Palladium	Pd	4.22	5.36	4.18	4.81	−21.3	1.0	−12.3
Platinum	Pt	6.74‡	7.58	6.31	6.66	−11.1	6.8	1.2
Plutonium	Pu	5.95	4.49	3.33	5.36	32.5	79	11.0
Rhenium	Re	7.9	9.91	7.09	8.52	−20	11	−7.3
Rhodium	Rh	5††	6.08	4.87	5.53	−18	2.7	−9.6
Ruthenium	Ru	6.1	6.62	5.21	5.96	−7.9	17	2.3
Silver	Ag	3.88	4.05	3.20	3.86	−4.2	21.3	0.5
Tantalum	Ta	8.6	8.33	5.63	7.22	3.2	53	19
Titanium	Ti	4.42	3.39	3.07	4.55	30.4	44.0	−2.9
Tungsten	W	7.0	9.85	6.03	8.20	−29	16	−15
Uranium	U	6.57	5.73	7.47	6.65	14.7	−12.0	−1.2
Zirconium	Zr	‡‡	3.95	3.63	3.85	–	–	–
					$\Delta(22)$ %	16.7	27.8	8.5
					$S(22)$	0.194	0.391	0.109

† Two individual data points are given, namely, 5.2 and 5.0 mPa s. A mean value has been adopted.
‡ Another datum point also given, namely, 4.82 mPa s.
†† Also another datum point given, namely, 2.9 mPa s.
‡‡ Four individual data points, namely, 4.7, 4.83, 3.5, and 5.45 mPa s.

Table 15.4 *Values of the physical quantities or the parameters used for calculating viscosities.*

Metal		M kg mol^{-1}	Tm K	V_m 10^{-6} m^3 mol^{-1}	γ_m N m^{-1}	$\xi_T^{1/2}$
Chromium	Cr	0.05200	2180	8.27	1.710	*0.776*†
Cobalt	Co	0.05893	1768	7.59	1.900	0.878
Copper	Cu	0.06355	1357.8	7.943	1.320	0.887
Gold	Au	0.19697	1337.3	11.35	1.150	1.175
Hafnium	Hf	0.17849	2506	16.1	1.614	*1.073*
Iridium	Ir	0.19222	2719	9.61	2.250	*1.024*
Iron	Fe	0.05585	1811	7.94	1.880	0.880
Manganese	Mn	0.05494	1519	9.54	1.152	0.767
Molybdenum	Mo	0.09594	2896	10.3	2.080	0.977
Nickel	Ni	0.05869	1728	7.425	1.795	0.889
Niobium	Nb	0.09291	2750	11.9	1.937	*1.015*
Palladium	Pd	0.10642	1828.1	10.14	1.482	*1.032*
Platinum	Pt	0.19508	2041.6	10.32	1.746	1.125
Plutonium	Pu	0.24406	913	14.66	0.550	0.737
Rhenium	Re	0.18621	3459	9.90	2.610	*0.987*
Rhodium	Rh	0.10291	2237	9.53	1.940	*1.045*
Ruthenium	Ru	0.10107	2607	9.27	2.260	*1.036*
Silver	Ag	0.10787	1234.9	11.6	0.926	0.983
Tantalum	Ta	0.18095	3290	12.1	2.050	0.924
Titanium	Ti	0.04788	1941	11.6	1.670	*0.800*
Tungsten	W	0.18384	3695	10.4	2.310	0.872
Uranium	U	0.23803	1408	13.78	1.552	*1.333*
Zirconium	Zr	0.09122	2128	16	1.500	*1.120*

† Numeric figures in italics denote values calculated using predicted sound velocities.

Table 15.3 compares experimental values for the melting point viscosity of the *d*-block liquid transition metals, plus plutonium and uranium (i.e. actinoid metals), with those calculated on the basis of the MFBG relation represented by Eq.(15.5), together with δ_i, Δ, and S values. For comparison, the results of calculations based on the Andrade and the FBG models are also shown in Table 15.3. The data used for the viscosity calculations are given in Table 15.4. As is obvious from Table 15.3, the MFBG relation performs excellently with the *d*-block liquid transition metals, plus the two liquid actinoid metals, giving $\Delta(22)$ and $S(22)$ values of 8.5 per cent and 0.109, respectively. This is very remarkable when compared with experimental uncertainties for these metals; it may be estimated that the calculated viscosities fall within the range of uncertainties associated with experimental measurements.

Figures 15.2 to 15.4 show comparisons of experimental viscosity values for the 23 liquid metals listed in Table 15.3, together with osmium and vanadium, against those calculated by the three models. In these figures, data points linked by a vertical line represent two or four different experimental values for a single metal. Except for the copper group, the iron group, and several other metals, the viscosities of liquid metals plotted in these figures are highly uncertain. Accurate and reliable experimental data are indispensable for the viscosities (in general, for any other thermophysical property, as well)

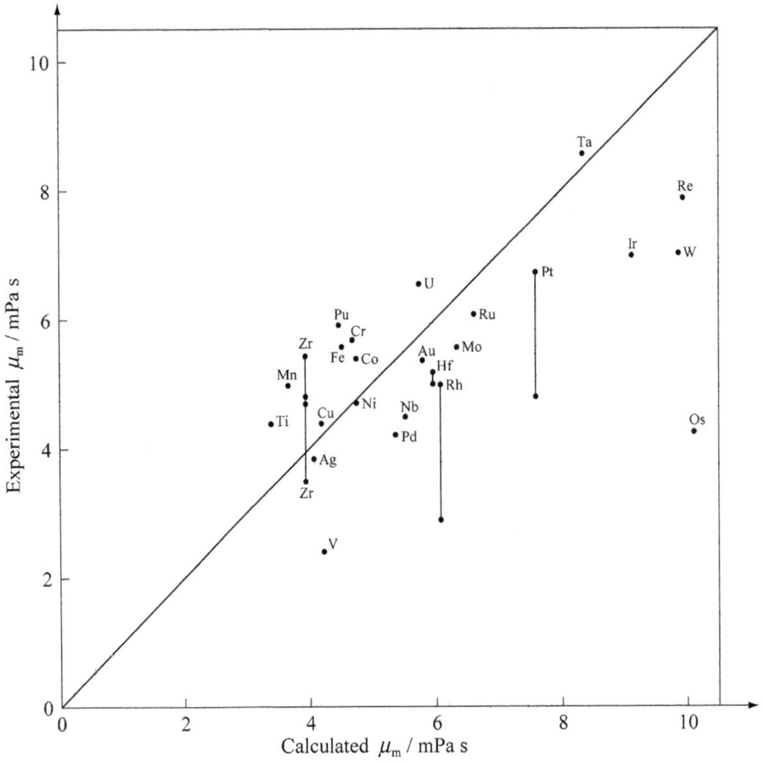

Figure 15.2 *Comparison of experimental values for the melting point viscosity of the d-block liquid transition metals plus plutonium and uranium (actinoid metals) with those calculated on the basis of the Andrade formula. Points linked by a vertical line represent two or four different experimental values for a single metal.*

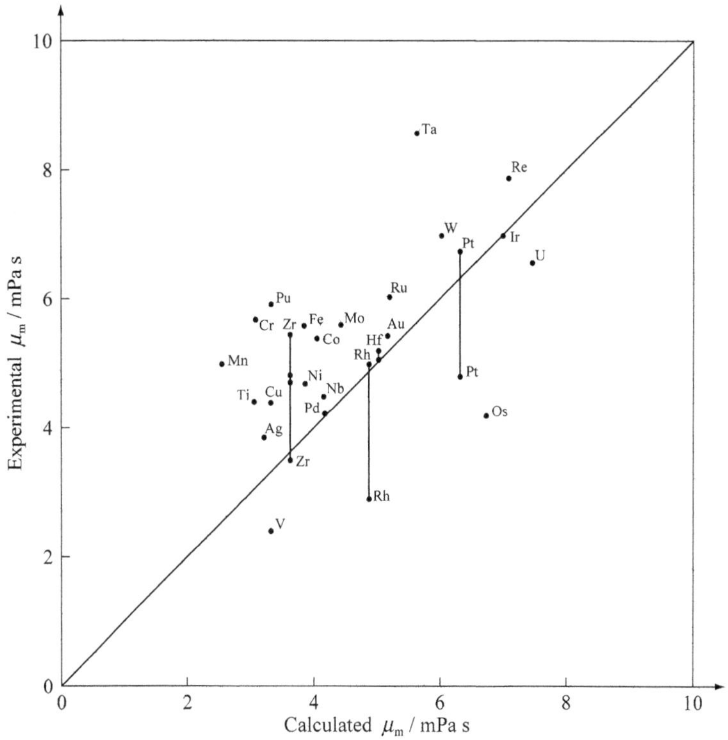

Figure 15.3 *Comparison of experimental values for the melting point viscosity of the d-block liquid transition metals plus plutonium and uranium (actinoid metals) with those calculated on the basis of the Fowler–Born–Green relation. Points linked by a vertical line represent two or four different experimental values for a single metal.*

of liquid metallic elements, if we are to devote time to subsequent studies. (A mixture of wheat and chaff gives rise to confusion.)

Values of the common parameter $\xi_T^{1/2}$ for some metals, shown in Table 15.3, were determined using predicted sound velocities. On the basis of Eqs.(15.2) and (15.4), the melting point viscosities of 12 liquid metals were calculated; in calculating the viscosities, predicted sound velocities were used for all 12 liquid metals. Table 15.5 gives a comparison between experimental and calculated viscosity values, together with δ_i, Δ, and S values. Also included in Table 15.5, for comparison, are their viscosities calculated on the basis of the Andrade and the FBG models. The data on k_0 and U_m used for the viscosity calculations are listed in Table 15.6. As is clear from Table 15.5, the MFBG relation provides excellent agreement with experimental values, in spite of using

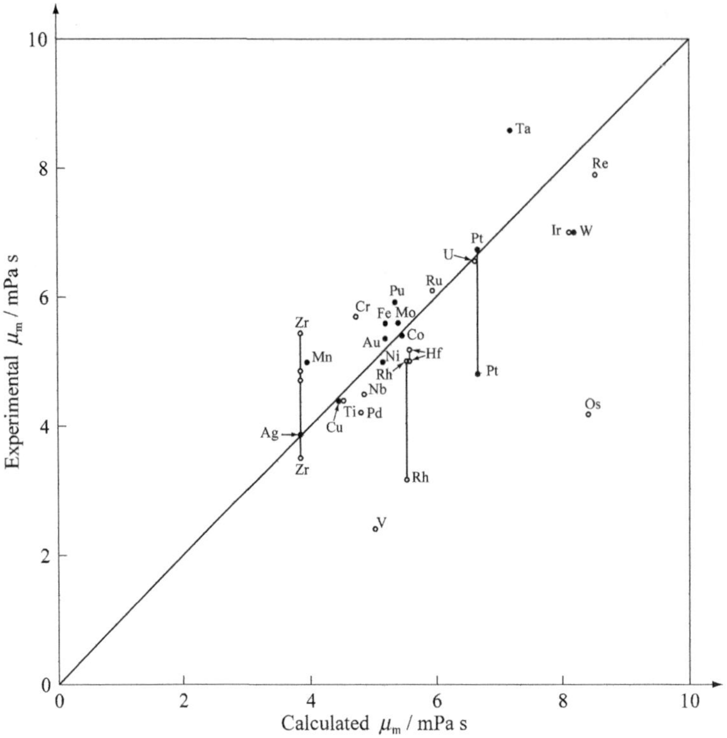

Figure 15.4 *Comparison of experimental values for the melting point viscosity of the d-block liquid transition metals, plus plutonium and uranium (actinoid metals) with those calculated on the basis of the modified Fowler–Born–Green relation:* ○*, calculated using predicted sound velocities (on the basis of Eq. (15.4) or (15.5)). Points linked by a vertical line represent two or four different experimental values for a single metal.*

both predicted sound velocity and numerical factor k_0 values. Incidentally, it is difficult to say which of the two (i.e. the MFBG(a) or the MFBG(b)) is better, in view of experimental uncertainties of ca. 15 per cent. Moreover, the results of viscosity calculations for liquid francium and uranium indicate that agreement with experimental data is surprisingly good, giving δ_{Fr} and δ_U values of 11.0 or 1.1 and 3.6 or −1.4 per cent (see Table 15.5), respectively. It can safely be said that values of the numerical factors k_0 (dimensionless), or C_0 (kg$^{1/2}$ m^{-1} s K$^{1/2}$ mol$^{1/2}$), for the *s*-block, and actinoid metals (probably, up to americium; see **Actinoids**) are, to a good approximation, 0.1105, or 0.341 (kg$^{1/2}$ m^{-1} s K$^{1/2}$ mol$^{1/2}$), and 0.0858, or 0.439 (kg$^{1/2}$ m^{-1} s K$^{1/2}$ mol$^{1/2}$), respectively. Thus, actinoid metals (probably, up to americium) may be included in the group of the *d*-block transition metals. In Figure 15.5, values of C_0 are plotted against atomic number.

Table 15.5 *Comparison of experimentally derived melting point viscosities of some liquid metals with those calculated on the basis of the modified Fowler–Born–Green relation, in which predicted sound velocities were used for the calculations, together with δ_i, Δ, and S values. For comparison, the results based on the Fowler–Born–Green and the Andrade models are also listed.*

Metal		$(\mu_m)_{exp}$ mPa s	$(\mu_m)_{cal}$ / mPa s				δ_i / %			
			MFBG(a)†	MFBG(b)‡	Andrade	FBG	MFBG(a)†	MFBG(b)‡	Andrade	FBG
Chromium	Cr	5.7	5.30	4.73	4.69	3.09	7.5	21	22	84
Francium	Fr	0.765	0.689	0.757	0.707	0.626	11.0	1.1	8.2	22.2
Hafnium	Hf	5.1	6.02	5.58	5.97	5.04	−15	−8.6	−15	1.2
Iridium	Ir	7.0	7.84	8.12	9.10	7.00	−11	−14	−23	0
Niobium	Nb	4.5	4.79	4.88	5.52	4.17	−6.1	−7.8	−18	7.9
Palladium	Pd	4.22	4.87	4.82	5.36	4.18	−13.3	−12.4	−21.3	1.0
Rhenium	Re	7.9	7.56	8.52	9.91	7.09	4.5	−7.3	−20	11
Rhodium	Rh	5	5.63	5.53	6.08	4.87	−11	−9.6	−18	2.7
Ruthenium	Ru	6.1	6.12	5.97	6.62	5.21	−0.3	2.2	−7.9	17
Titanium	Ti	4.42	4.17	4.56	3.39	3.07	6.0	−3.1	30.4	44.0
Uranium	U	6.57	6.34	6.66	5.73	7.47	3.6	−1.4	14.7	−12.0
Yttrium	Y	3.6	3.37	–	2.96	2.27	6.8	–	22	59
						Δ (12) %	8.0	8.0††	18.4	21.8
						S (12)	0.090	0.099‡‡	0.194	0.337

† k_0 values were determined through interpolation using a plot of k_0 against atomic number, which are given in Table 15.6.
‡ k_0 values of 0.0858 (i.e. $C_0 = 0.439$ kg$^{-1/2}$ m^{-1} s K$^{1/2}$ mol$^{1/2}$) and 0.1105 (i.e. $C_0 = 0.341$ kg$^{-1/2}$ m^{-1} s K$^{1/2}$ mol$^{1/2}$) were used for the *d*-block transition metals, plus uranium, and the *s*-block metal (i.e. francium), respectively.
†† $\Delta(11)$ / %.
‡‡ $S(11)$.

Table 15.6 *Values of the parameters used for calculating viscosities on the basis of the modified Fowler–Born–Green relation; the results of calculations are given in MFBG(a) (see Table 15.5).*

Metal		k_0	$U_m^{\dagger}$ m s^{-1}
Chromium	Cr	0.0766	4216
Francium‡	Fr	0.1214	741
Hafnium	Hf	0.0795	3371
Iridium	Ir	0.0888	3230
Niobium	Nb	0.0873	4629
Palladium	Pd	0.0849	3586
Rhenium	Re	0.0967	3569
Rhodium	Rh	0.0843	4087
Ruthenium	Ru	0.0837	4411
Titanium	Ti	0.0937	4270
Uranium	U	0.0900	2718
Yttrium‡	Y	0.0968	2677

† Predicted values (see Chapter 11).
‡ Melting point surface tensions, $(\gamma_m)_{exp}$, for francium and yttrium used for the calculations are 0.062 and 0.872 N m^{-1}, respectively.

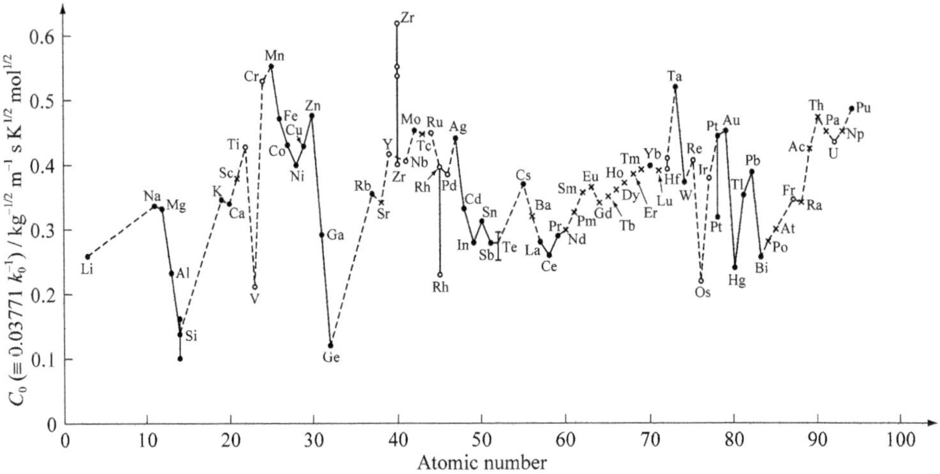

Figure 15.5 *Plot of the numerical factor C_0 ($\equiv 0.03771k_0^{-1}$) appearing in Eq. (7.41) against atomic number: ●, I, calculated using experimental data; ○, calculated using predicted sound velocities; ×, calculated using both predicted viscosities and sound velocities, except for barium and strontium; calculated using predicted viscosities and experimental sound velocities only for barium and strontium. Data points linked by a vertical line indicate values calculated from two, three, or four different experimental viscosities for a single metallic element.*

15.4 Viscosity Calculations using Periodic Relationships and the Weakness of this Method

We give a few examples of viscosity calculations. Values of both C_A and k_0 are determined through interpolation using a plot of C_A or k_0 against atomic number Z (see Figures 15.1 and 15.6, and Table 15.2).

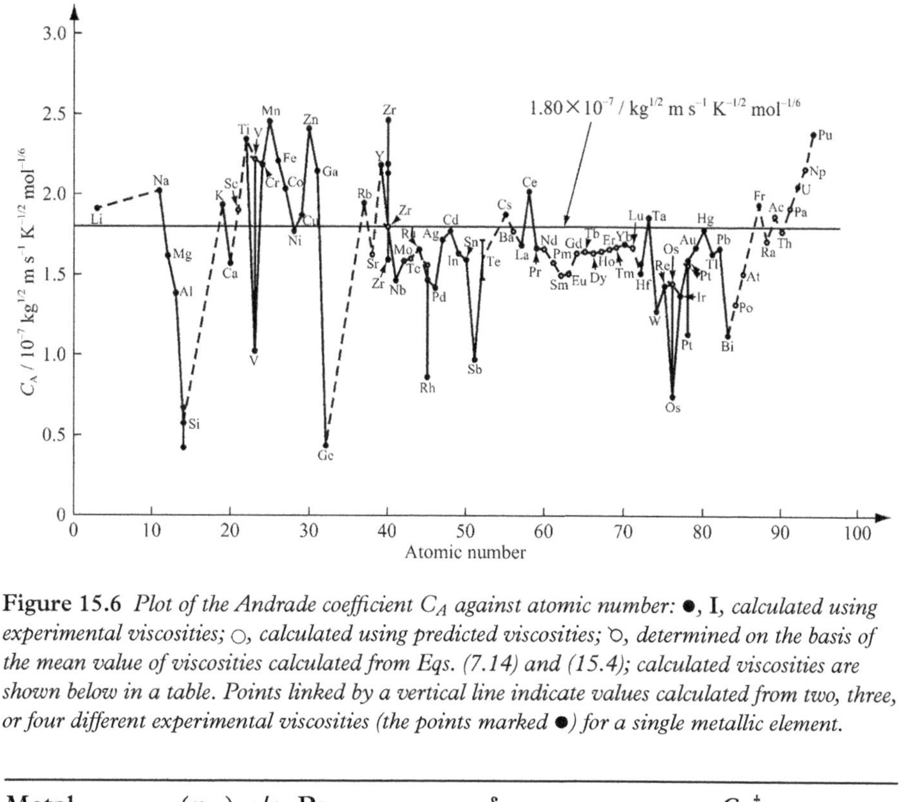

Figure 15.6 *Plot of the Andrade coefficient C_A against atomic number: ●, I, calculated using experimental viscosities; ○, calculated using predicted viscosities; ↄ, determined on the basis of the mean value of viscosities calculated from Eqs. (7.14) and (15.4); calculated viscosities are shown below in a table. Points linked by a vertical line indicate values calculated from two, three, or four different experimental viscosities (the points marked ●) for a single metallic element.*

Metal	$(\mu_m)_{cal}$/mPa s			δ_{diff} %	C_A† 10^{-7}kg$^{1/2}$ ms^{-1} K$^{-1/2}$ mol$^{-1/6}$
	Eq. (7.14)	Eq. (15.4)	Mean		
Os	7.91	8.39	8.15	-5.9	1.41
Pt	6.44	6.67	6.56	-3.5	1.53
Rh	5.20	5.53	5.37	-6.1	1.54
V	5.34	5.02	5.18	6.2	2.27
Zr	4.02	3.85	3.94	4.3	1.83

† Obtained through interpolation using a periodic relationship (see Table 15.2).
The other data used for the viscosity calculations are given in Chapter 17.

(1) Calculation of the melting point viscosity of liquid cadmium:

Z	Metal	C_A 10^{-7}kg$^{1/2}$ m s^{-1} K$^{-1/2}$ mol$^{-1/6}$	k_0
47	Ag	1.72	0.0855
48	Cd	*1.68*	*0.1102*
49	In	1.64	0.1348

i) Based on the Andrade-type model
Data for M, T_m, V_m, γ_m, and U_m are listed in Chapter 17.

M / kg mol^{-1} $\quad$ T_m/K

$$(\mu_m)_{cal} = 1.68 \times 10^{-7} \frac{(0.1124 \times 594.2)^{1/2}}{(14.0 \times 10^{-6})^{2/3}} = 2.36 \times 10^{-3} / \text{Pa s} (= 2.36 / \text{mPa s})$$

C_A / kg$^{1/2}$ m s^{-1} K$^{-1/2}$ mol$^{-1/6}$ $\quad$ V_m / m^3 mol^{-1}

(predicted value)

ii) Based on the MFBG relation

γ_m / N m^{-1}

$$(\mu_m)_{cal} = \frac{0.637}{0.1102 \times 2237} = 2.58 \times 10^{-3} / \text{Pa s} (= 2.58 / \text{mPa s})$$

k_0 $\quad$ U_m / m s^{-1}

(predicted value)

(2) Calculation of the melting point viscosity of cerium:

Z	Metal	C_A 10^{-7}kg$^{1/2}$ m s^{-1} K$^{-1/2}$ mol$^{-1/6}$	k_0
57	La	1.69	0.1354
58	Ce	*1.69*	*0.1330*
		(1.685)	
59	Pr	1.68	0.1305

i) Based on the Andrade-type model

$$(\mu_m)_{cal} = 2.72\,\text{mPa s}$$
$$(2.71\,\text{mPa s})$$

ii) Based on the MFBG relation

$$(\mu_m)_{cal} = 3.53\,\text{mPa s}$$

(3) Calculation of the melting point viscosity of thallium:

Z	Metal	C_A $10^{-7}\text{kg}^{1/2}\,\text{m s}^{-1}\,\text{K}^{-1/2}\,\text{mol}^{-1/6}$	k_0
80	Hg	1.79	0.1580
81	Tl	*1.74*	*0.1275*
82	Pb	1.68	0.0969

i) Based on the Andrade-type model

$$(\mu_m)_{cal} = 2.75\,\text{mPa s}$$

ii) Based on the MFBG relation

$$(\mu_m)_{cal} = 2.19\,\text{mPa s}$$

The results of these calculations and values of δ_i are listed in the table.

Metal		$(\mu_m)_{exp}$ mPa s	$(\mu_m)_{cal}$ / mPa s			δ_{diff}† %	δ_i / %		
			Andrade-type	MFBG	Mean		Andrade-type	MFBG	Mean
Cadmium	Cd	2.50	2.36	2.58	2.47	−8.9	5.9	−3.1	1.2
Cerium	Ce	3.25	2.72	3.53	3.13	−25.9	19.5	−7.9	3.8
Thallium	Tl	2.59	2.75	2.19	2.47	22.7	−5.8	18.3	4.9

† The relative difference in the values calculated using the two models, δ_{diff} (%), is evaluated by

$$\delta_{diff}(\%) = \frac{(\mu_m)_{cal\ \text{Andrade-type}} - (\mu_m)_{cal\ \text{M F-B-G}}}{(\mu_m)_{cal}\ \text{(i.e Mean value)}} \times 100$$

In the method of interpolation based on a periodic relationship, it is difficult to accurately predict a value of any thermophysical property if there is a sharp variation in its value against atomic number. This is the weakness of the interpolation method. In such cases, a fairly (or very) large discrepancy exists, in general, between values calculated using two methods, as shown in the viscosity calculations of liquid cerium and thallium. Nevertheless, this provides a chance to consider the problem of the large discrepancy, particularly if the relative differences δ_{diff} in values calculated using two models is more than approximately 30 per cent, although the mean value determined from two models (e.g. the Andrade-type model, and the MFBG relation) often provides a better result, as already indicated in the above examples.

In the case of a metallic liquid's viscosity, fairly large discrepancies exist among experimental values (see, for example, Figures 7.21 to 7.25). Viscosities for the majority of liquid metals have yet to be established. For example, the melting point viscosity $(\mu_{\text{m}})_{\text{exp}}$ of liquid aluminium may be probably lie in the range 1.1–1.4 mPa s, i.e. 1.25 $\pm$ 0.15 mPa s, or $\pm$12 per cent around the mean (see Figure 7.23); the $\delta_{\text{exp,diff}}$ value, evaluated by the following equation, for the melting point viscosity of liquid aluminium is 24 per cent:

$$\delta_{\text{exp diff}} = \frac{(\mu_{\text{m}})_{\text{exp Max}} - (\mu_{\text{m}})_{\text{exp Min}}}{\overline{(\mu_{\text{m}})}_{\text{exp}}} = \frac{1.4 - 1.1}{1.25} \times 100 = 24\ (\%)$$

where $(\mu_{\text{m}})_{\text{exp Max}}$ and $(\mu_{\text{m}})_{\text{exp Min}}$ are the extremes of reliable values for $(\mu_{\text{m}})_{\text{exp}}$, respectively, and $\overline{(\mu_{\text{m}})}_{\text{exp}} = \left\{(\mu_{\text{m}})_{\text{exp Max}} + (\mu_{\text{m}})_{\text{exp Min}}\right\}/2$.

Since liquid cerium shows sharp changes in values of various thermophysical properties plotted against atomic number (e.g. melting point temperature, evaporation enthalpy, sound velocity, surface tension), similar behaviour can be estimated for its viscosity value. Basically, this may depend on the electronic state, or the electron configuration, in the metallic liquid. In the case of thallium, owing to the inert-pair effect (see **Inert-pair effect**), the thallium in the liquid state would be univalent, which has approximately the same k_0 value of 0.1105 as the *s*-block metals. Using the k_0 value of 0.1105 for thallium, calculated values of μ_{m} and δ_{Tl} are 2.52 mPa s and 2.8 per cent, respectively; we see that the calculated value of the melting point viscosity of liquid thallium coincides well with experimental value.

15.5 Calculations or Predictions of Melting Point Viscosity

At present, experimentally derived viscosity data on liquid metallic elements are available for 53 or so. Viscosities for some rare earth and radioactive metals (e.g. technetium, radium, thorium) have yet to be measured experimentally. The melting point viscosities of the liquid metals are calculated on the basis of the Andrade-type and the MFBG models.

To be specific, the present authors' approach to the calculations of the melting point viscosities of liquid metals is based on the following approximations or assumptions.

Table 15.7 *Calculated, or predicted, melting point viscosities of 22 liquid metals.*

Metal		$(\mu_m)_{cal}$ / mPa s		δ_{diff}† %	Adopt. $(\mu_m)_{cal}$‡ mPa s
		Eq. (7.14)	Eq. (15.1)		
Actinium*	Ac	2.42	2.58	−6.4	2.50
Astatine*	At	1.13	1.08	4.5	1.11
Barium	Ba	1.75	1.73	1.1	1.74
Dysprosium	Dy	3.87	3.65	5.9	3.76
Erbium	Er	4.16	3.95	5.2	4.06
Europium	Eu	2.28	1.79	24.0	2.04
Gadolinium	Gd	3.60	3.38	6.3	3.49
Holmium	Ho	4.00	3.81	4.9	3.91
Lutetium	Lu	4.36	4.63	−6.0	4.50
Neodymium	Nd	2.96	2.88	2.7	2.92
Neptunium*	Np	5.56	5.28	5.2	5.42
Polonium*	Po	1.22	1.17	4.2	1.20
Promethium*	Pm	3.22	2.83	12.9	3.03
Protactinium*	Pa	6.35	6.02	5.3	6.19
Radium*	Ra	1.71	1.48	14.4	1.48
Samarium	Sm	3.23	2.51	25.1	2.87
Scandium	Sc	2.70	2.54	6.1	2.62
Strontium	Sr	1.79	1.41	23.8	1.41
Technetium*	Tc	5.48	5.31	3.3	5.40
Terbium	Tb	3.73	3.54	5.2	3.64
Thorium*	Th	5.20	4.52	14.0	4.86
Thulium	Tm	3.75	3.65	2.7	3.70

† The relative difference in values calculated from Eqs. (7.14) and (15.1).
‡ Adopted values of calculated viscosities.
* Radioactive metals.

(i) Basically, both C_A and k_0 values are determined on the basis of the interpolation method using a C_A or k_0 plot *vs.* atomic number Z. Incidentally, uncertainties in the experimental data used are estimated to be about 1 to 20 per cent (depending on kinds of metallic elements). (ii) Values of C_A vary periodically with atomic number Z, the periodic Group IA metals occupying the peaks of the C_A vs. Z plot; in other words, C_A

Table 15.8 *Values of the physical quantities or parameters used for calculating viscosities.*

Metal		C_A 10^{-7} $kg^{1/2}$ $m\ s^{-1}$ $K^{-1/2}$ $mol^{-1/6}$	M $kg\ mol^{-1}$	T_m K	$V_m^{2/3}$ 10^{-4} $m^2\ mol^{-2/3}$	k_0	γ_m $N\ m^{-1}$	U_m $m\ s^{-1}$
Actinium*	Ac	*1.80*†	0.22703	1323	*12.91*†	*0.0858*†	*0.427*†	*1929*†
Astatine*	At	*1.54*	0.20999	575	*14.96*	*0.1280*	*0.190*	*1370*
Barium	Ba	*1.79*	0.13733	1000	11.96	*0.1189*	0.273	1331
Dysprosium	Dy	*1.69*	0.16250	1685	7.22	*0.1073*	0.648	*1654*
Erbium	Er	*1.70*	0.16726	1802	7.095	*0.1007*	0.637	*1603*
Europium	Eu	*1.69*	0.15197	1095	9.569	*0.1173*	0.264	*1257*
Gadolinium	Gd	*1.69*	0.15725	1586	7.417	*0.1140*	0.664	*1725*
Holmium	Ho	*1.69*	0.16493	1747	7.170	*0.1040*	0.650	*1639*
Lutetium	Lu	*1.62*	0.17497	1936	6.843	*0.0940*	0.940	*2162*
Neodymium	Nd	*1.68*	0.14424	1294	7.749	*0.1272*	0.687	*1873*
Neptunium*	Np	*2.23*	0.23705	903	*5.869*	*0.0858*	*0.768*	*1694*
Polonium*	Po	*1.34*	0.20898	527	*11.56*	*0.1372*	*0.235*	*1464*
Promethium*	Pm	*1.68*	0.14692	1443	*7.596*	*0.1239*	*0.606*	*1727*
Protactinium*	Pa	*1.97*	0.23104	2113	*6.853*	*0.0858*	*1.559*	*3020*
Radium	Ra	*1.97*	0.22603	973	*17.08*	*0.1105*	*0.244*	*1491*
Samarium	Sm	*1.69*	0.15036	1347	7.442	*0.1206*	0.430	*1422*
Scandium	Sc	*1.96*	0.04496	1814	6.560	*0.1023*	0.939	*3608*
Strontium	Sr	*2.07*	0.08762	1050	11.10	*0.1105*	0.296	1902
Technetium*	Tc	*1.63*	0.09691	2443	*4.576*	*0.0858*	*1.930*	*4240*
Terbium	Tb	*1.69*	0.15893	1629	7.294	*0.1106*	0.669	*1709*
Thorium*	Th	*1.89*	0.23204	2023	7.875	*0.0858*†	1.006	*2594*
Thulium	Tm	*1.70*	0.16893	1818	*7.946*	*0.0974*	*0.581*	*1634*

* Radioactive.

† Numerical figures in italics denote values predicted by the authors in this book.

Table 15.9 *Adopted viscosities of the twenty-two liquid metallic elements, together with values for their numerical factors C_A, k_0, C_0, and k_v (cf. Figures 15.1, 15.5, 15.6, and 15.9).*

Metal		$(\mu_m)_{cal}$ † 10^{-3}Pa s	C_A 10^{-7} kg$^{1/2}$ ms^{-1} K$^{-1/2}$ mol$^{-1/6}$	k_0	C_0 kg$^{-1/2}$ m^{-1} s K$^{1/2}$ mol$^{1/2}$	k_v
Actinium	Ac	2.50	1.86	0.0885	0.426	1.206
Astatine	At	1.11	1.51	0.1249	0.302	0.825
Barium	Ba	1.74	1.78	0.1179	0.320	1.468
Dysprosium	Dy	3.76	1.64	0.1042	0.362	1.595
Erbium	Er	4.06	1.66	0.0979	0.385	1.786
Europium	Eu	2.04	1.51	0.1030	0.366	1.771
Gadolinium	Gd	3.49	1.64	0.1103	0.342	1.428
Holmium	Ho	3.91	1.65	0.1014	0.372	1.671
Lutetium	Lu	4.50	1.67	0.0966	0.390	1.360
Neodymium	Nd	2.92	1.66	0.1256	0.300	1.087
Neptunium	Np	5.42	2.17	0.0836	0.451	1.176
Polonium	Po	1.20	1.32	0.1338	0.282	0.692
Promethium	Pm	3.03	1.58	0.1158	0.326	1.277
Protactinium	Pa	6.19	1.92	0.0834	0.452	1.025
Radium	Ra	1.48	1.71	0.1105	0.341	1.074
Samarium	Sm	2.87	1.50	0.1054	0.358	1.705
Scandium	Sc	2.62	1.90	0.0993	0.380	1.513
Strontium	Sr	1.41	1.63	0.1105	0.341	1.408
Technetium	Tc	5.40	1.61	0.0843	0.447	1.200
Terbium	Tb	3.64	1.65	0.1075	0.351	1.487
Thorium	Th	4.86	1.77	0.0798	0.473	1.218
Thulium	Tm	3.70	1.68	0.0961	0.392	1.783

† Adopted viscosity values.

values of the Group IIA metals are smaller than those of the Group IA metals. (iii) Values of k_0 for the *s*-block metals (excluding lithium) is 0.1105, on average. (iv) Values of k_0 for the *d*-block transition metals and actinoid metals is 0.0858, on average. (v) Viscosities of the *s*-block, and the *d*-block transition metals, plus actinoid metals can be calculated from Eqs.(15.2) or (15.3), and (15.4) or (15.5), respectively. (vi) The mean value of calculated viscosities using the two models is adopted, excluding radium and strontium.

Table 15.7 lists calculated or predicted values for the melting point viscosity of 22 liquid metals. The data used for the calculations are shown in Table 15.8. Table 15.9 gives adopted viscosities of the 22 liquid metallic elements, and their values for the numerical factors, appearing in Eqs.(7.14), (6.37), or (15.1), and (7.41).

The probable uncertainties in the calculated viscosity values, listed in Table 15.7, are estimated to be 5 to 25 per cent.

Figure 15.7 includes a plot of both experimental and calculated values of μ_m against atomic number, and shows the periodic variation of μ_m. Since viscosity is a dynamic, or non-equilibrium, property, the trend in variation of viscosity with increase in the atomic number of the metallic element is different from equilibrium properties, such as melting point temperature, and surface tension (cf. Figures 10.2 and 14.6). Thus, owing to the effect of atomic mass (or molar mass), based on atomic motions, actinoid metals have slightly higher viscosity values compared with those the *d*-block transition metals in periods 4 and5.

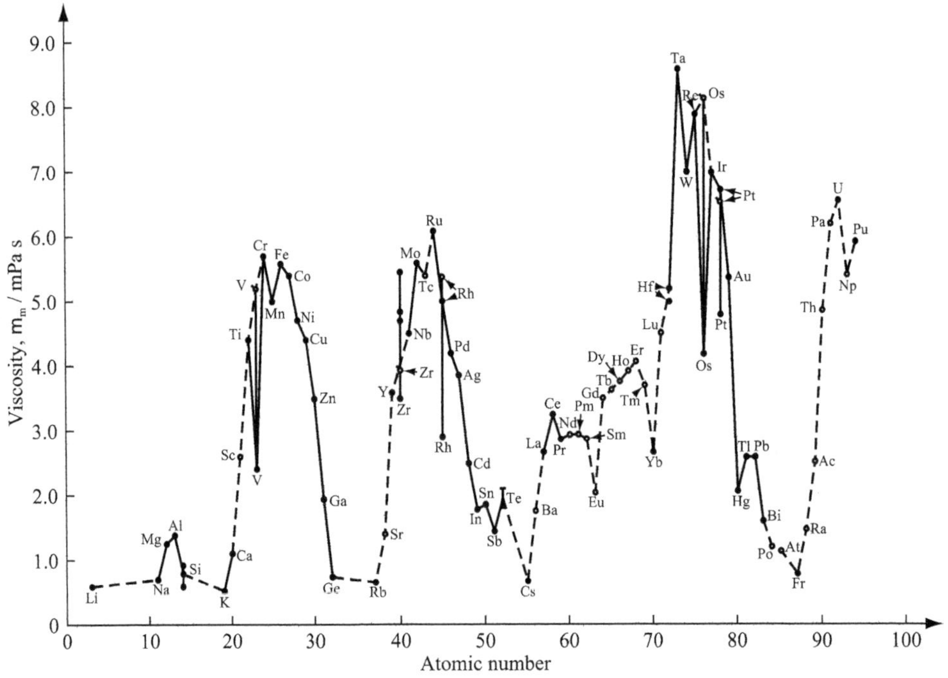

Figure 15.7 *Periodic variation of the melting point viscosities with increase in the atomic number of the metallic element: ●, I, experimental values; ○, calculated, or predicted, values; ꝺ, the mean values of viscosities calculated from Eqs. (7.14) and (15.4). Points linked by a vertical line represent two, three, or four different experimental values for a single metallic element.*

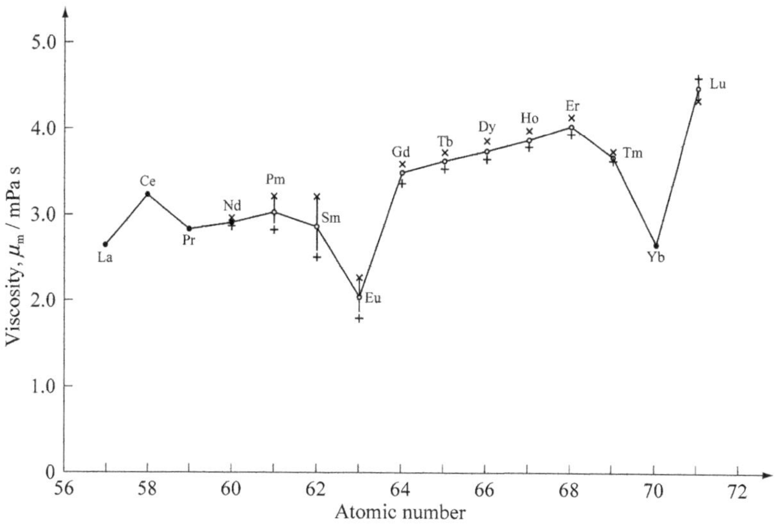

Figure 15.8 *Plot of the melting point viscosities of liquid lanthanoid metals vs. atomic number: ●, experimental values; ×, calculated values on the basis of the Andrade-type model, i.e. Eq. (7.14); +, calculated values on the basis of the modified Fowler–Born–Green relation (or model), i.e. Eq. (7.41) or (15.1); ○, the mean values of viscosities calculated on the basis of the two models.*

In Figure 15.8, the periodicity of the melting point viscosity of liquid lanthanoid metals is shown. Overall, their viscosity values increase with increasing atomic number. However, europium and ytterbium, which are divalent in the metallic state, have local minima in the μ_m vs. Z curve.

15.6 Supplementary Explanation for the Modified Fowler–Born–Green Relation

We now give a supplementary explanation for the MFBG relation. At the melting point temperature, the FBG relation may be written as

$$\mu_m = \frac{16}{15} k_v \left(\frac{m}{kT_m} \right)^{1/2} \gamma_m = \frac{16}{15 R^{1/2}} k_v \left(\frac{M}{T_m} \right)^{1/2} \gamma_m \tag{15.6}$$

where k_v is a dimensionless parameter which is introduced so as to exactly reproduce the respective experimental viscosity values of liquid metallic elements. We can rewrite Eq.(15.6) in the form

$$k_v = \frac{15}{16} \left(\frac{kT_m}{m} \right)^{1/2} \frac{\mu_m}{\gamma_m} \tag{15.6'}$$

Values of k_v are also given in Tables 15.2 and 15.9. On the other hand, combination of Eqs.(5.32) and (15.1) leads to

$$\mu_m = \frac{1}{9.197 R^{1/2} k_0 \xi_T^{1/2}} \left(\frac{M}{T_m} \right)^{1/2} \gamma_m \tag{15.7}$$

From Eqs.(15.6) and (15.7), we have

$$k_v = \frac{0.1019}{k_0 \xi_T^{1/2}} \tag{15.8}$$

Since the value of k_0 is roughly equal to 0.10, we have

$$k_v \simeq \frac{1}{\xi_T^{1/2}} \qquad (15.9)^4$$

A comparison between Eqs.(10.6) and (15.8) gives

$$k_v = Q^{-1} = C^{-1}, \text{ or } k_v Q = k_v C = 1 \tag{15.10}$$

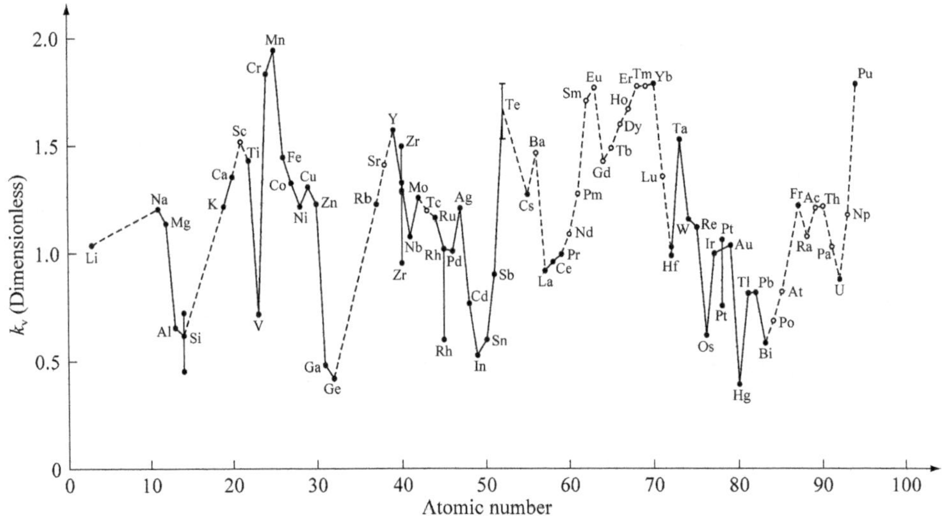

Figure 15.9 *Plot of the parameter k_v appearing in Eq. (15.6) against atomic number:* ●, **I**, *calculated using experimental viscosities;* ○, *calculated using predicted viscosities. Points linked by a vertical line represent values calculated from two, three, or four different experimental viscosity values for a single metallic element.*

[4] We have already mentioned that the mean values of k_0 for the *s*-block and the *d*-block liquid metals at their melting point temperatures are 0.1105 and 0.0858, respectively, so that we obtain $k_v = 0.922/\xi_T^{1/2}$ for the *s*-block liquid metals, and $k_v = 1.19/\xi_T^{1/2}$ for the *d*-block liquid transition metals.

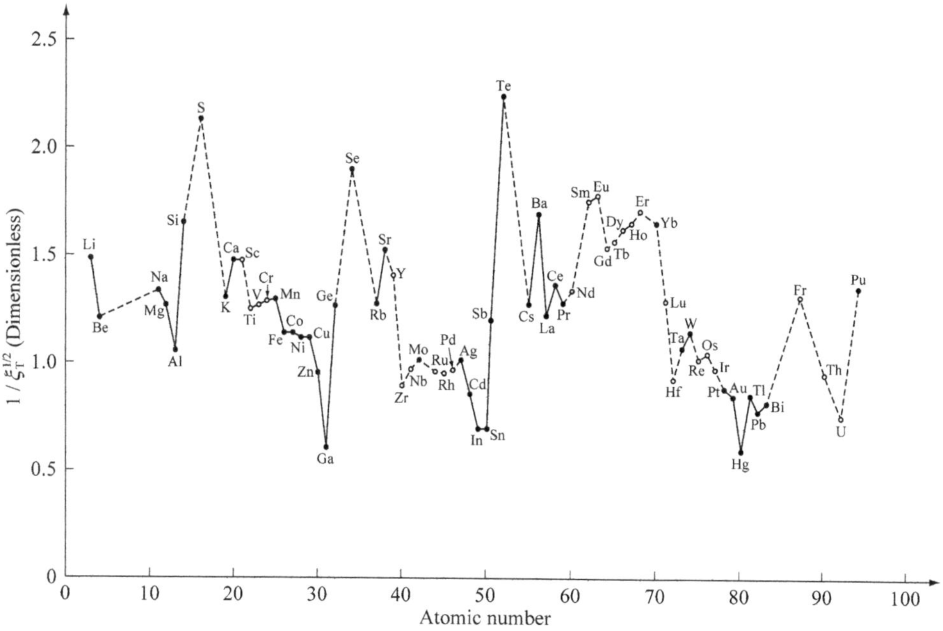

Figure 15.10 *Plot of the reciprocal of the common parameter $1/\xi_T^{1/2}$ vs. atomic number:* ●, *calculated using experimental sound velocities;* ○, *calculated from predicted sound velocities.*

Figure 15.9 shows values of k_v plotted against atomic number Z. We see that the Group IIIA or IVA metallic elements, excluding thallium and lead, and mercury occupy the valleys of the plot. Figure 15.10 gives a plot of $1/\xi_T^{1/2}$vs. Z. Roughly speaking, in the plot of Figure 15.10 a similar trend to that of Figure 15.9 is captured, particularly, if silicon, germanium, selenium, and sulphur (non-metal) are excluded because they have abnormally low (silicon and germanium) or abnormally high (selenium) viscosity values. Unfortunately, the current accuracy of experimental viscosity data is insufficient for detailed discussions. For example, in Figure 15.9, the Group IB metals (i.e. copper group metals) may occupy the minor peaks and the Group IVB metals the minor valleys of the plot. Experimental uncertainties of ca. 5 per cent are needed for even more development of materials process science.

15.7 Temperature Dependence of Viscosity

We showed in Chapter 7 that the temperature dependence of the viscosity of liquid metallic elements is given by

$$\mu = \mu_0 \exp\left(\frac{H_\mu}{RT}\right) \tag{7.47}$$

$$\mu_0 = \mu_m \left(0.279 T_m^{0.29}\right)^{-1}$$

$$H_\mu = 2.32 T_m^{1.29}$$

This model for the variation of viscosity with temperature performs reasonably well for the 30 liquid metallic elements listed in Table 7.8 or plotted in Figure 7.7, giving $\Delta(30)$ and $S(30)$ values of 22.6 per cent and 0.289, respectively; however, these results are not completely satisfactory.

This viscosity model is expressed in terms of only one parameter of the melting point temperature T_m. Since (comparatively) accurate values of the melting point temperature are already known for all liquid metallic elements, in view of practicability, the model represented by Eq.(7.47) may be endowed with the necessary conditions (i.e. both accuracy and universality) of being predictive. Unfortunately, most of experimentally derived data for the temperature dependence of metallic liquids are not very reliable, because of difficulties encountered in obtaining accurate viscosity values at high temperatures. Needless to say, any model must be checked on the basis of accurate and reliable experimental data.

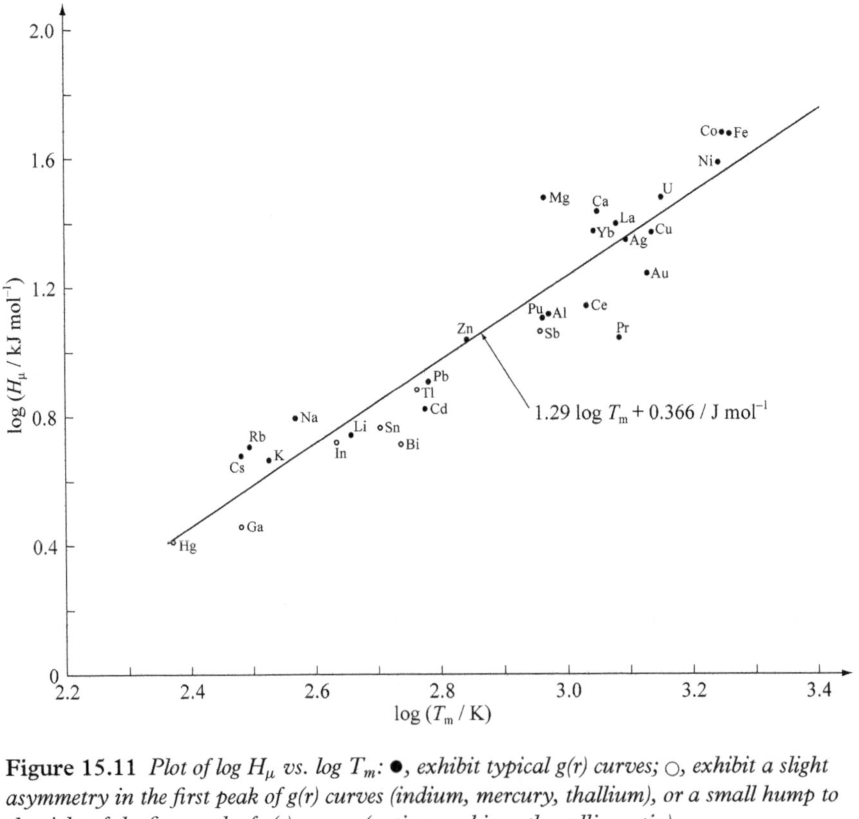

Figure 15.11 *Plot of log H_μ vs. log T_m: ●, exhibit typical g(r) curves; ○, exhibit a slight asymmetry in the first peak of g(r) curves (indium, mercury, thallium), or a small hump to the right of the first peak of g(r) curves (antimony, bismuth, gallium, tin).*

Figure 15.11 shows a plot of $\log H_\mu$ against $\log T_m$ for the 30 liquid metallic elements. As mentioned in Chapter 2, several liquid metallic elements constitute the characteristic features of the pair distribution functions obtained from experimental scattering data, namely, they display a slight asymmetry in the first peak, or a small hump to the right of the first peak. In Figure 15.11, these metallic elements are plotted with open circles. As is seen from the plot of this figure, all of these metallic elements exist below the straight line of $\log H_\mu = 1.29 \log T_m + 0.366$, which was determined so as to give the minimum S value for the 30 liquid metallic elements; in other words, all of them have negative δ_i values (i.e. $\delta_{Hg} = -1.3$ to $\delta_{Bi} = -34.0$ per cent). Thus the data points can be divided into two groups, as indicated in Figure 15.11. It is expected that this reasonable classification may improve the performance of the model. We repeat here that accurate experimental data for the viscosity of metallic liquids are needed for the development of materials process science.

REFERENCES

1. T. Iida, R. Guthrie, M. Isac, and N. Tripathi, *Metall. Mater. Trans. B*, 37 (2006), 403.
2. P.-F. Paradis, T. Ishikawa, and N. Koike, *Int. J. Refract. Met. Hard Mater.*, 25 (2007), 95.
3. T. Ishikawa and P.-F. Paradis, in *High-Temperature Measurements of Materials*, H. Fukuyama and Y. Waseda (eds.), Springer, Berlin, 2009, p.190.
4. P.-F. Paradis, T. Ishikawa, and S. Yoda, *Int. J. Thermophys.*, 23 (2002), 825.

16

Predictions of Self-Diffusivity

16.1 Introduction

The most characteristic feature of all metallic liquids is their capacity to readily flow. This means the high mobility of individual atoms in the liquid state. Thus numerous theoretical, semi-theoretical, or semi-empirical as well as experimental studies of self-diffusion and solute diffusion in metallic liquids have been made over the past century.

To study atomic diffusion phenomena in a liquid alloy system, accurate data for the self-diffusivities of the respective pure metal components of the system are first needed. Nevertheless, experimental data for self-diffusivities in liquid metallic elements are scanty. At present, it would appear that experimentally derived self-diffusivity data are available only for some 20 liquid metallic elements. Consequently, in Chapter 8, the authors proposed a predictive model, expressed in terms of a few well-known physical quantities, for self-diffusivity in the liquid metallic elements, from the standpoint of materials process science.

In this chapter, liquid metallic self-diffusivity data, i.e. liquid metallic self-diffusivities at the melting point temperatures, and their temperature dependence, are calculated or predicted for a large number of liquid metallic elements.

16.2 A Predictive Model for Liquid Metallic Self-Diffusivity

16.2.1 Melting Point Self-Diffusivity of Liquid Metallic Elements

Various models and equations for self-diffusion in liquid metals have been proposed, as described in Chapter 8. Of these, the modified Stokes–Einstein formula is the most satisfactory from the viewpoint of materials process science, because it is simple and is in good agreement with self-diffusivities for liquid metallic elements.

On the basis of the modified Stokes–Einstein formula, that was discussed in Chapter 8, the melting point self-diffusivity D_m of liquid metallic elements can be obtained through a predictive model represented by the following equation, comprising three macroscopic physical quantities:

The Thermophysical Properties of Metallic Liquids: Volume 2 – Predictive Models. First Edition.
Takamichi Iida and Roderick I. L. Guthrie.
Published in 2015 by Oxford University Press.

$$D_m = 2.19 \times 10^{-16} \frac{T_m}{\mu_m V_m^{1/3}} \tag{8.31}$$

These three physical quantities, i.e. the melting point temperature T_m, viscosity μ_m, and molar volume V_m, are already known for a great number of liquid metallic elements, as mentioned in the previous chapter. We showed in Chapter 8 that Eq. (8.31) performs extremely well for liquid metallic elements, giving $\Delta(18)$ and $S(18)$ values of 7.8 per cent and 0.102, respectively; most importantly, the self-diffusivity values calculated from Eq. (8.31) fall within the range of uncertainties associated with experimental measurements (experimental uncertainties in self-diffusivity data are estimated to be about 5 to 25 per cent).

Incidentally, Eq. (8.31) can also be written in the form (cf. Chapter 8).

$$D_m = 2.19 \times 10^{-16} \left(\frac{\xi_T T_m}{M} \right)^{1/2} \frac{T_m}{C_0 V_m^{1/3} \gamma_m} \tag{8.30}$$

16.2.2 Temperature Dependence of Liquid Metallic Self-Diffusivity

Convective flows can have significant effects on the accuracy of liquid diffusivity measurements. Thus, measurements of liquid metallic diffusivity under microgravity have been carried out in order to avoid natural convection, produced by gravity, in liquid diffusion experiments.

Figure 16.1 shows self-diffusivity values of liquid tin, measured under microgravity, as a function of temperature [1]. As can be seen, however, even for the self-diffusivity of liquid tin, there are still considerable uncertainties in the experimental data; the discrepancy between the experimental values of liquid tin, obtained by the two research groups, is ca. 15 per cent, and further, the temperature dependence also appears to be inconsistent between the two data sets.

The Arrhenius-type equation normally holds, to a good approximation, for liquid metallic self-diffusivity

$$D = D_0 \exp\left(-\frac{H_D}{RT} \right) \tag{8.32}$$

$$H_D = 13.5\ T_m^{1.10}, \text{ or } \log H_D = 1.10\ \log T_m + 1.13 \tag{8.34}$$

$$D_0 = D_m \exp\left(\frac{H_D}{RT_m} \right) = D_m \exp\left(1.62\ T_m^{0.10}\right)$$

We repeat here that the Arrhenius-type equation is adopted only as a mathematical expression for describing the self-diffusivity of liquid metallic elements.

Although liquid metallic self-diffusivity data are sparse, we can readily predict values for the self-diffusivity of liquid metallic elements using data for their melting point

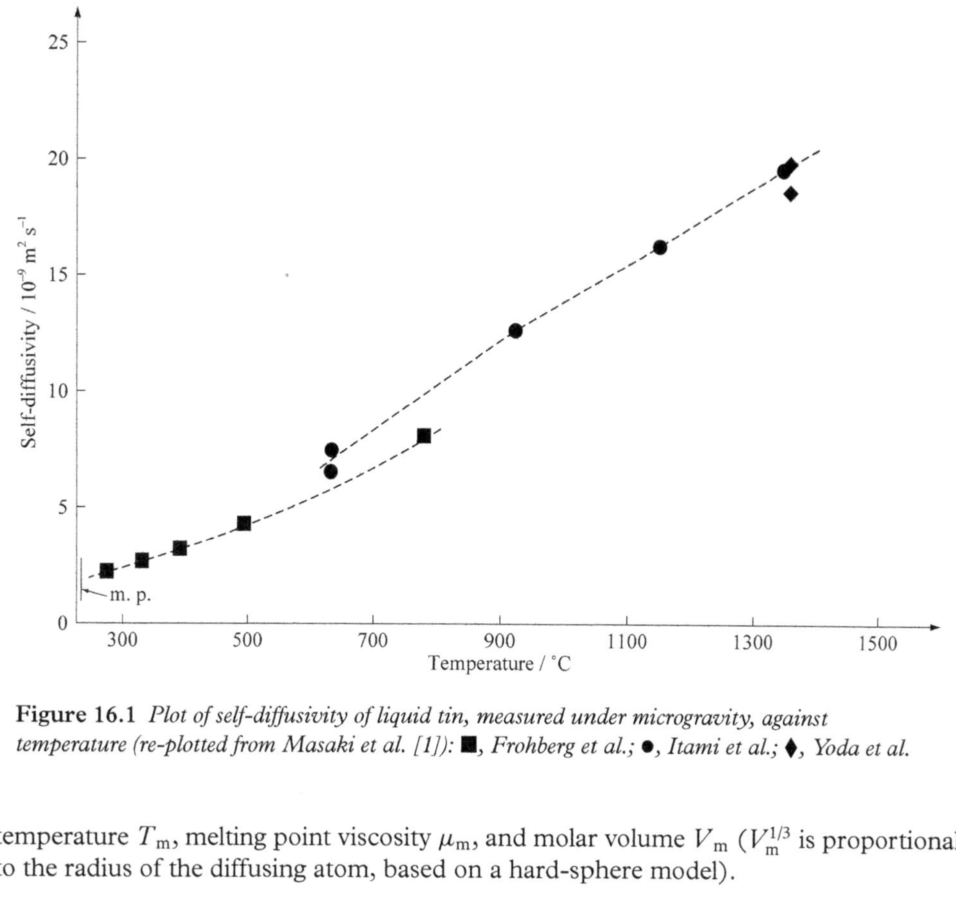

Figure 16.1 *Plot of self-diffusivity of liquid tin, measured under microgravity, against temperature (re-plotted from Masaki et al. [1]): ■, Frohberg et al.; ●, Itami et al.; ♦, Yoda et al.*

temperature T_m, melting point viscosity μ_m, and molar volume V_m ($V_m^{1/3}$ is proportional to the radius of the diffusing atom, based on a hard-sphere model).

16.3 Calculations or Predictions of Liquid Metallic Self-Diffusivity Data

We now calculate, or predict, melting point self-diffusivity data, i.e. D_m and H_D, for a large number of liquid metallic elements (excluding the 18 liquid metallic elements listed in Table 8.6). On the basis of Eqs. (8.31) and (8.34), the self-diffusivity data are calculated for 57 liquid metallic elements. The results of calculations are given in Table 16.1, together with the physical quantities used for calculating their self-diffusivity data. The calculated, or predicted, values are expected to fall, or almost fall, within the range of uncertainties of any experimental measurement made in the future.

Figure 16.2 shows a plot of the melting point self-diffusivity values for a large number of liquid metallic elements against atomic number. Roughly speaking, the liquid

Table 16.1 *Calculated (or predicted) self-diffusivity data for liquid metallic elements, together with values of the physical quantities used for the calculations.*

(a)

Element		$(D_m)_{cal}$ 10^{-9} m^2 s^{-1}	H_D kJ mol^{-1}	μ_m 10^{-3} Pa s	T_m K	$V_m^{1/3}$ 10^{-2} m mol$^{-1/3}$
Actinium*	Ac	3.23	36.6	*2.50*†	1323	*3.593*†
Aluminium	Al	6.60	25.0	1.38	933.5	2.245
Astatine*	At	2.93	14.6	*1.11*	575	*3.867*
Barium	Ba	3.64	26.9	*1.74*	1000	3.458
Calcium	Ca	7.20	30.3	1.10	1115	3.085
Cerium	Ce	2.62	29.0	3.25	1071	2.757
Chromium	Cr	4.14	63.4	5.7	2180	2.022
Cobalt	Co	3.65	50.4	5.4	1768	1.965
Dysprosium	Dy	3.65	47.8	*3.76*	1685	2.687
Erbium	Er	3.65	51.4	*4.06*	1802	2.664
Europium	Eu	3.80	29.7	*2.04*	1095	3.093
Francium	Fr	1.88	7.16	0.765	300	4.561
Gadolinium	Gd	3.65	44.7	*3.49*	1586	2.723
Gold	Au	2.43	37.1	5.37	1337.3	2.247
Hafnium	Hf	4.18	73.9	5.2	2506	2.525
		4.35		5.0		
Holmium	Ho	3.65	49.7	*3.91*	1747	2.678
Iridium	Ir	4.00	80.9	7.0	2719	2.126
Iron	Fe	3.55	51.7	5.6	1811	1.995
Lanthanum	La	3.43	32.6	2.66	1191	2.857
Lutetium	Lu	3.60	55.7	*4.50*	1936	2.616
Magnesium	Mg	6.52	24.6	1.25	923	2.482
Manganese	Mn	3.14	42.6	5	1519	2.121
Molybdenum	Mo	5.20	86.7	5.6	2896	2.176
Neodymium	Nd	3.49	35.7	*2.92*	1294	2.784

Table 16.1 *(continued)*

(a)

Element		$(D_m)_{cal}$ 10^{-9} m^2 s^{-1}	H_D kJ mol^{-1}	μ_m 10^{-3} Pa s	T_m K	$V_m^{1/3}$ 10^{-2} m mol$^{-1/3}$
Neptunium*	Np	1.51	24.1	*5.42*	903	*2.423*
Nickel	Ni	4.13	49.1	4.7	1728	1.951
Niobium	Nb	5.86	81.9	4.5	2750	2.283
Palladium	Pd	4.38	52.3	4.22	1828.1	2.164
Platinum	Pt	4.26	59.0	4.82	2041.6	2.177
		3.05		6.74		
		3.13		6.56‡		
Plutonium*	Pu	1.37	24.4	5.95	913	2.447
Polonium*	Po	2.83	13.3	*1.20*	527	*3.400*
Praseodymium	Pr	3.34	33.0	2.85	1204	2.772
Promethium*	Pm	3.78	40.3	*3.03*	1443	*2.756*
Protactinium*	Pa	2.86	61.3	*6.19*	2113	*2.618*
Radium*	Ra	3.48	26.1	*1.48*	973	*4.133*
Rhenium	Re	4.47	105.4	7.9	3459	2.147
Rhodium	Rh	7.97	65.3	2.9	2237	2.120
		4.62		5		
		4.30		5.37‡		
Ruthenium	Ru	4.45	77.2	6.1	2607	2.101
Samarium	Sm	3.77	37.4	*2.87*	1347	2.728
Scandium	Sc	5.92	51.8	*2.62*	1814	2.561
Selenium	Se	0.161	12.4	24.8	493.7	2.701
Strontium	Sr	4.89	28.4	*1.41*	1050	3.332
Tantalum	Ta	3.65	99.8	8.6	3290	2.296
Technetium*	Tc	4.63	71.9	*5.40*	2443	*2.139*
Terbium	Tb	3.63	46.0	*3.64*	1629	2.701
Thorium*	Th	3.25	58.4	*4.86*	2023	2.806

continued

Table 16.1 *(continued)*

(a)

Element		$(D_m)_{cal}$ 10^{-9} m^2 s^{-1}	H_D kJ mol^{-1}	μ_m 10^{-3} Pa s	T_m K	$V_m^{1/3}$ 10^{-2} m mol$^{-1/3}$
Thulium	Tm	3.82	51.9	*3.70*	1818	*2.819*
Titanium	Ti	4.25	55.8	4.42	1941	2.264
Tungsten	W	5.30	113.3	7.0	3695	2.183
Uranium*	U	1.96	39.2	6.57	1408	2.397
Ytterbium	Yb	3.03	29.7	2.67	1092	2.955
Yttrium	Y	3.94	51.2	3.6	1795	2.772
Zirconium	Zr	3.93	61.8	4.7	2128	2.520
		3.83		4.83		
		5.28		3.5		
		3.39		5.45		
		4.69		3.94‡		

(b)

Element		D_m 10^{-9} m^2 s^{-1}		H_D kJ mol^{-1}	μ_m 10^{-3} Pa s	T_m K	$V_m^{1/3}$ 10^{-2} m mol$^{-1/3}$
		Exp	Cal				
Bismuth	Bi	3.78	2.73	13.8	1.59	544.6	2.750
		0.8002					
Osmium	Os		8.15	100.3	4.2	3306	2.115
			4.20		8.15‡		
Silicon	Si	40 ± 5 [2]	28.5	47.8	0.58	1685	2.231
			20.7		0.8		
			17.6		0.94		
Vanadium	V		9.41	63.5	2.4	2183	2.118
			4.36		5.18‡		

† Numerical figures in italics denote predicted values (see Tables 15.8 and 15.9).
‡ Calculated values (see Figure 15.7).
* Radioactive.

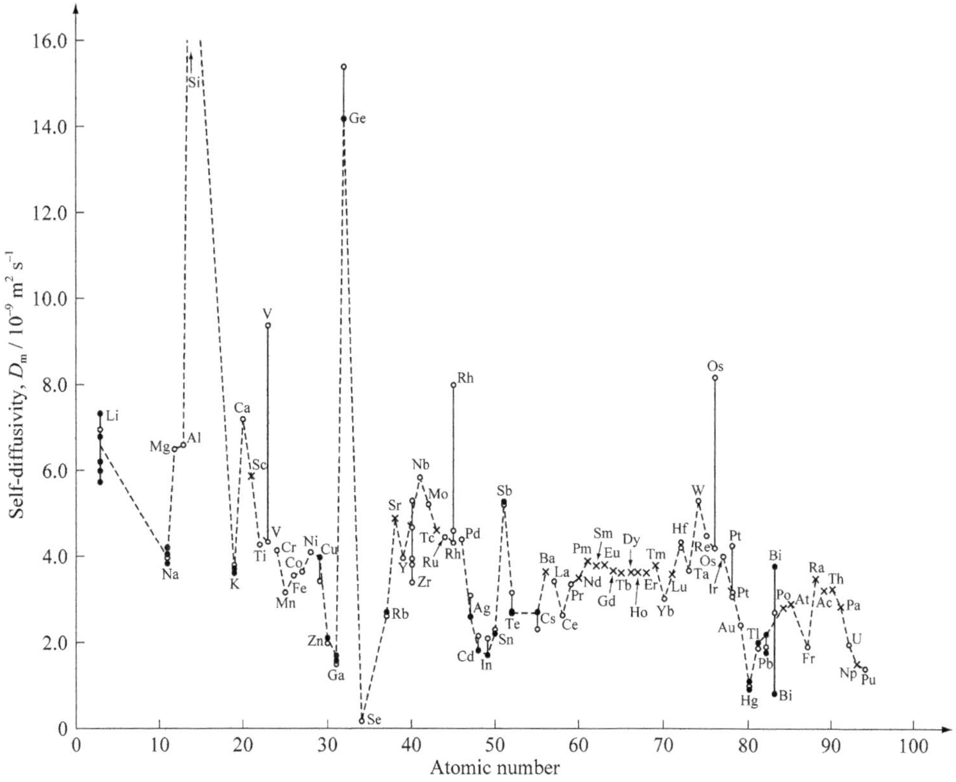

Figure 16.2 *Variation of melting point self-diffusivity values for 75 liquid metallic elements with atomic number: ●, experimental; ○, calculated, or predicted, using experimental viscosities; ̀○, calculated using calculated viscosities (see Figure 15.7); ×, calculated using predicted viscosities (see Table 15.9).*

self-diffusivity values tend to decrease with increase in the atomic number of the metallic element. The Group IIA metals in the periodic table may occupy the peaks of the plot (a slight difference in the self-diffusivity values between magnesium and aluminium is within the range of their uncertainties). To conclude, accurate and reliable experimental data for liquid metallic self-diffusivities are needed to check the predictive model.

REFERENCES

1. T. Masaki, T. Fukazawa, S. Matsumoto, T. Itami, and S. Yoda, *Meas. Sci. Technol.*, **16** (2005), 327.
2. P.G. Sanders and M.J. Aziz, *J. Appl. Phys.*, **86** (1999), 4258.

17

Experimentally Derived Data for the Thermophysical Properties of Liquid Metallic Elements

This chapter provides a compilation of experimental data for the thermophysical properties of liquid metallic elements. The properties covered include melting and boiling points (Table 17.1), atomic numbers and relative atomic masses (Table 17.2), molar melting and evaporation enthalpies (Table 17.3), densities and related data (Table 17.4), vapour pressure equations (Table 17.5), molar heat capacities at constant pressure (Table 17.6), sound velocity data (Table 17.7), surface tension data (Table 17.8), viscosity data (Table 17.9), self-diffusivity data (Table 17.10), and electrical resistivity and thermal conductivity data (Table 17.11).

17.1 Melting and Boiling Points

Table 17.1 *Melting and boiling points.*

Element		Melting point† K	Boiling point† K
Actinium	Ac	1323[a]	3473±300[a]
Aluminium	Al	933.5‡	2792
Americium	Am	1267±4[a]	2880[a]
Antimony	Sb	903.8‡	1860
Arsenic	As	1090 at 2.8 MPa[a]	886 sublimation[a]
Astatine	At	575[b]	610[b]
Barium	Ba	1000	2170
Beryllium	Be	1560	2744

continued

The Thermophysical Properties of Metallic Liquids: Volume 2 – Predictive Models. First Edition.
Takamichi Iida and Roderick I. L. Guthrie.
Published in 2015 by Oxford University Press.

Table 17.1 *(continued)*

Element		Melting point[†] K	Boiling point[†] K
Bismuth	Bi	554.6	1837
Boron	B	2573[a]	2823[a]
Cadmium	Cd	594.2[‡]	1040
Caesium	Cs	301.6	944
Calcium	Ca	1115	1757
Carbon	C	sublimation at 3925[a]	
Cerium	Ce	1071	3706
Chromium	Cr	2180	2944
Cobalt	Co	1768[‡]	3200
Copper	Cu	1357.8[‡]	2835
Dysprosium	Dy	1685	2840
Erbium	Er	1802	3141
Europium	Eu	1095[a]	1800[a]
Francium	Fr	300[b]	950[b]
Gadolinium	Gd	1586	3546
Gallium	Ga	302.9[‡]	2477
Germanium	Ge	1210	3103
Gold	Au	1337.3[‡]	3129
Hafnium	Hf	2506	4876
Holmium	Ho	1747	2973
Indium	In	429.7[‡]	2345
Iridium	Ir	2719[‡]	4701
Iron	Fe	1811	3134
Lanthanum	La	1191	3737
Lead	Pb	600.6	2022
Lithium	Li	453.7	1615
Lutetium	Lu	1936	3675

Table 17.1 *(continued)*

Element		Melting point† K	Boiling point† K
Magnesium	Mg	923	1363
Manganese	Mn	1519	2334
Mercury	Hg	234.3	629.9
Molybdenum	Mo	2896	4912
Neodymium	Nd	1294	3347
Neptunium	Np	903±1[a]	–
Nickel	Ni	1728‡	3186
Niobium	Nb	2750	5017
Osmium	Os	3306	5285
Palladium	Pd	1828.1	3236
Phosphorus	P	317.3[a]	553[a]
(yellow)		863 at 4.3 MPa (red)[a]	
Platinum	Pt	2041.6	4098
Plutonium	Pu	913	3501
Polonium	Po	527	1235
Potassium	K	336.5	1032
Praseodymium	Pr	1204	3793
Promethium	Pm	1443[b]	2733[b]
Protactinium	Pa	2113[b]	–
Radium	Ra	973[b]	1413[b]
Rhenium	Re	3459	5869
Rhodium	Rh	2237	3968
Rubidium	Rb	312.5	961
Ruthenium	Ru	2607	4423
Samarium	Sm	1347	2067
Scandium	Sc	1814	3109
Selenium	Se	493.7	958

continued

Table 17.1 *(continued)*

Element		Melting point† K	Boiling point† K
Silicon	Si	1685	3543
Silver	Ag	1234.9‡	2435
Sodium	Na	370.9	1156
Strontium	Sr	1050	1655
Sulphur	S	386.0[a]	717.8[a]
(rhombohedral)		392.2[a] (monoclinic)	–
Tantalum	Ta	3290	5731
Technetium	Tc	2443[b]	–
Tellurium	Te	723	1261
Terbium	Tb	1629	3503
Thallium	Tl	577	1746
Thorium	Th	2023	5061
Thulium	Tm	1818	2223
Tin	Sn	505.1‡	2875
Titanium	Ti	2941	3560
Tungsten	W	3695	5828
Uranium	U	1408	4404
Vanadium	V	2183	3680
Ytterbium	Yb	1092	1469
Yttrium	Y	1795	3618
Zinc	Zn	692.7‡	1180
Zirconium	Zr	2128	4682

† Under standard pressure (see Table 1.1).
‡ Defined fixed point ITS-90: The International Temperature Scale of 1990 (see Table 1.1).
Data, except for those bearing the superscripts a and b, are taken from Gale and Tolemeier [1].
[a] Benenson et al. [2].
[b] Nagakura et al. [3].

17.2 Atomic Numbers and Relative Atomic Masses

Table 17.2 *Atomic numbers and relative atomic masses.*

Element		Atomic number	Relative atomic mass
Actinium	Ac	89	227.0278*
Aluminium	Al	13	26.981539
Americium	Am	95	243.0614*
Antimony	Sb	51	121.757
Argon	Ar	18	39.948
Arsenic	As	33	74.92159
Astatine	At	85	209.9871*
Barium	Ba	56	137.327
Berkelium	Bk	97	247.0703*
Beryllium	Be	4	9.012182
Bismuth	Bi	83	208.98037
Boron	B	5	10.811
Bromine	Br	35	79.904
Cadmium	Cd	48	112.411
Caesium	Cs	55	132.90543
Calcium	Ca	20	40.078
Californium	Cf	98	251.0796*
Carbon	C	6	12.011
Cerium	Ce	58	140.115
Chlorine	Cl	17	35.4527
Chromium	Cr	24	51.9961
Cobalt	Co	27	58.93320
Copper	Cu	29	63.546
Curium	Cm	96	247.0703*
Dysprosium	Dy	66	162.50
Einsteinium	Es	99	252.0829*

continued

Table 17.2 *(continued)*

Element		Atomic number	Relative atomic mass
Erbium	Er	68	167.26
Europium	Eu	63	151.965
Fermium	Fm	100	257.0951*
Fluorine	F	9	18.998403
Francium	Fr	87	223.0197*
Gadolinium	Gd	64	157.25
Gallium	Ga	31	69.723
Germanium	Ge	32	72.61
Gold	Au	79	196.96654
Hafnium	Hf	72	178.49
Helium	He	2	4.002602
Holmium	Ho	67	164.93032
Hydrogen	H	1	1.00794
Indium	In	49	114.818
Iodine	I	53	126.90447
Iridium	Ir	77	192.22
Iron	Fe	26	55.847
Krypton	Kr	36	83.80
Lanthanum	La	57	138.9055
Lawrencium	Lr	103	260.1053*
Lead	Pb	82	207.2
Lithium	Li	3	6.941
Lutetium	Lu	71	174.967
Magnesium	Mg	12	24.3050
Manganese	Mn	25	54.93805
Mendelevium	Md	101	258.0986*
Mercury	Hg	80	200.59

Table 17.2 *(continued)*

Element		Atomic number	Relative atomic mass
Molybdenum	Mo	42	95.94
Neodymium	Nd	60	144.24
Neon	Ne	10	20.1797
Neptunium	Np	93	237.0482*
Nickel	Ni	28	58.6934
Niobium	Nb	41	92.90638
Nitrogen	N	7	14.00674
Nobelium	No	102	259.1009*
Osmium	Os	76	190.23
Oxygen	O	8	15.9994
Palladium	Pd	46	106.42
Phosphorus	P	15	30.973762
Platinum	Pt	78	195.08
Plutonium	Pu	94	244.0642*
Polonium	Po	84	208.9824*
Potassium	K	19	39.0983
Praseodymium	Pr	59	140.90765
Promethium	Pm	61	146.9151*
Protactinium	Pa	91	231.0358*
Radium	Ra	88	226.0254*
Radon	Rn	86	222.0176*
Rhenium	Re	75	186.207
Rhodium	Rh	45	102.90550
Rubidium	Rb	37	85.4678
Ruthenium	Ru	44	101.07
Samarium	Sm	62	150.36
Scandium	Sc	21	44.955910

continued

Table 17.2 *(continued)*

Element		Atomic number	Relative atomic mass
Selenium	Se	34	78.96
Silicon	Si	14	28.0855
Silver	Ag	47	107.8682
Sodium	Na	11	22.989768
Strontium	Sr	38	87.62
Sulphur	S	16	32.066
Tantalum	Ta	73	180.9479
Technetium	Tc	43	96.9063*
Tellurium	Te	52	127.60
Terbium	Tb	65	158.92534
Thallium	Tl	81	204.3833
Thorium	Th	90	232.0381*
Thulium	Tm	69	168.93421
Tin	Sn	50	118.710
Titanium	Ti	22	47.88
Tungsten	W	74	183.84
Uranium	U	92	238.0289*
Vanadium	V	23	50.9415
Xenon	Xe	54	131.29
Ytterbium	Yb	70	173.04
Yttrium	Y	39	88.90585
Zinc	Zn	30	65.39
Zirconium	Zr	40	91.224

* Radioactive.

17.3 Molar Melting and Evaporation Enthalpies

Table 17.3 *Molar melting and evaporation enthalpies.*

Element		Melting enthalpy $\Delta_l^g H_m$ / kJ mol^{-1}	Evap.enthalpy $\Delta_l^g H_b$ / kJ mol^{-1}
Aluminium	Al	10.7	294
Antimony	Sb	20.3	165[a]
Barium	Ba	7.7	151
Beryllium	Be	12.5	294
Bismuth	Bi	10.9	152
Boron	B	22.2	541
Cadmium	Cd	6.3	100
Caesium	Cs	2.18	65.9
Calcium	Ca	8.66	150
Cerium	Ce	5.5	314
Chromium	Cr	14.6	348
Cobalt	Co	16.2	383
Copper	Cu	13.0	304
Dysprosium	Dy	11.1	251[b]
Erbium	Er	19.9	271[b]
Europium	Eu	9.21	176[b]
Gadolinium	Gd	10.0	312[b]
Gallium	Ga	5.63	254
Germanium	Ge	34.7[c]	333[c]
Gold	Au	12.9	325
Hafnium	Hf	26.1	571[a]
Holmium	Ho	17.0	–
Indium	In	3.27	226
Iridium	Ir	22.5	564[b]
Iron	Fe	15.5	354
Lanthanum	La	11.3	400

continued

Table 17.3 *(continued)*

Element		Melting enthalpy $\Delta_{\mathrm{l}}^{\mathrm{g}}H_{\mathrm{m}}$ / kJ mol^{-1}	Evap.enthalpy $\Delta_{\mathrm{l}}^{\mathrm{g}}H_{\mathrm{b}}$ / kJ mol^{-1}
Lead	Pb	4.77[c]	180[c]
Lithium	Li	4.19	142
Lutetium	Lu	22.0	–
Magnesium	Mg	8.94	132
Manganese	Mn	14.6	230
Mercury	Hg	2.37	57.2
Molybdenum	Mo	27.8	590[a]
Neodymium	Nd	7.14	284[b]
Nickel	Ni	17.8	380
Niobium	Nb	31.0	696
Osmium	Os	29.3[b]	–
Palladium	Pd	16.7	361[a]
Platinum	Pt	21.7	469[a]
Plutonium	Pu	2.80[a]	344[a]
Polonium	Po	12.5[b]	106[b]
Potassium	K	2.33	77.4
Praseodymium	Pr	6.89	333[b]
Rhenium	Re	33.1	707
Rhodium	Rh	22.4	496[b]
Rubidium	Rb	2.20	75.2
Ruthenium	Ru	19.5	568[b]
Samarium	Sm	8.62	192[b]
Scandium	Sc	14.1	305
Selenium	Se	5.42	94.8
Silicon	Si	50.2[c]	395
Silver	Ag	11.3	253
Sodium	Na	2.63[c]	89.1[c]

Table 17.3 *(continued)*

Element		Melting enthalpy $\Delta_l^g H_m$ / kJ mol^{-1}	Evap.enthalpy $\Delta_l^g H_b$ / kJ mol^{-1}
Strontium	Sr	8.24	139
Sulphur	S	1.35	9.30
Tantalum	Ta	36.0	753
Tellurium	Te	17.5[c]	–
Terbium	Tb	10.8	293[b]
Thallium	Tl	4.21	162
Thorium	Th	13.8	544
Thulium	Tm	16.7	–
Tin	Sn	7.06	291
Titanium	Ti	15.5	430
Tungsten	W	35.3	800
Uranium	U	8.71	423[b]
Vanadium	V	23.0	458
Ytterbium	Yb	7.67	–
Yttrium	Y	11.4	393
Zinc	Zn	7.26	115
Zirconium	Zr	20.0	582

Data, except for those bearing the superscripts a, b, or c, are taken from Benenson et al. [2].

[a] Iida and Guthrie [4].

[b] Japan Institute of Metals [5].

[c] Nagakura et al. [3].

17.4 Density Data

Table 17.4 (a) *Density data.*

Element		T_m °C	ρ_m 10^3 kg m^{-3}	$-\Lambda^{\dagger}$ 10^{-1} kg m^{-3} K^{-1}	Range‡	V_m 10^{-6} m^3 mol^{-1}	$\alpha_m^{\dagger\dagger}$ 10^{-4} K^{-1}
Aluminium	Al	660.3	2.385	3.5	2.4–4.0	11.31	1.5
			2.380[a]	3.5[a]	–	11.34	1.5
Antimony	Sb	630.6	6.483	8.2	4.5–11.8	18.78	1.3
Arsenic	As	817 (at 2.8 MPa)	5.22	5.4	–	14.4	1.0
Barium	Ba	727	3.321	2.7	–	41.35	0.81
Beryllium	Be	1287	1.690	1.2	–	5.333	0.71
Bismuth	Bi	271.4	10.05	11.8	10.8–14.1	20.79	1.17
Boron	B	2300	2.08	–	–	5.20	–
Cadmium	Cd	321.1	8.01	12.2	11.4–14.1	14.0	1.52
Caesium	Cs	28.4	1.84	5.7	5.5–6.0	72.2	3.1
Calcium	Ca	842	1.365	2.2	–	29.36	1.6
Cerium	Ce	798	6.685	2.3	–	20.96	0.34
			6.55[b]	7.0[b]	–	21.4	1.1
Chromium	Cr	1907	6.29	7.2	–	8.27	1.1
Cobalt	Co	1495	7.76	10.9	9.5–12.5	7.59	1.40
			7.750[a]	11[a]	–	7.604	1.4
Copper*	Cu	1084.6	8.000	8.0	7.2–10.0	7.943	1.0
			7.960[a]	7.6[a]	–	7.983	0.95
Dysprosium	Dy	1412	8.37[b]	14.3[b]	–	19.4	1.71
Erbium	Er	1529	8.86[b]	15.7[b]	–	18.9	1.77
Europium	Eu	822	5.13[b]	28[b]	–	29.6	5.5
Francium	Fr	27	2.35	7.92	–	94.9	3.37
Gadolinium	Gd	1313	7.79[b]	10[b]	–	20.2	1.3
			7.404[b]	4.44[b]	–	21.24	0.600
Gallium	Ga	29.8	6.10[c]	5.6	5.1–7.3	11.4	0.92

Table 17.4 (a) *(continued)*

Element		T_m °C	ρ_m 10^3 kg m^{-3}	$-\Lambda^{\dagger}$ 10^{-1} kg m^{-3} K^{-1}	Range[‡]	V_m 10^{-6} m^3 mol^{-1}	$\alpha_m^{\dagger\dagger}$ 10^{-4} K^{-1}
Germanium	Ge	937	5.49	4.9	4.7–4.9	13.2	0.89
Gold*	Au	1064.2	17.36	15	–	11.35	0.86
Hafnium	Hf	2233	11.1	–	–	16.1	–
Holmium	Ho	1474	8.58[b]	14.7[b]	–	19.2	1.71
			8.293[b]	4.2[b]	–	19.89	0.51
Indium	In	156.6	7.03[c]	6.8	6.8–9.4	16.3	0.97
Iridium	Ir	2446	20.0	–	–	9.61	–
Iron	Fe	1538	7.03	8.8	7.3–9.6	7.94	1.3
			7.030[a]	8.6[a]	–	7.944	1.2
Lanthanum	La	918	5.955	2.4	–	23.33	0.40
			5.94[b]	6.1[b]	–	23.4	1.0
Lead	Pb	327.5	10.678	13.2	12.0–13.3	19.404	1.24
Lithium	Li	180.5	0.518	1.0	–	13.4	1.9
Lutetium	Lu	1663	9.75[b]	20.4[b]	–	17.9	2.09
Magnesium	Mg	650	1.590	2.6	–	15.29	1.6
			1.590[a]	2.6[a]	–	15.29	1.6
Manganese	Mn	1246	5.76	9.2	–	9.54	1.6
Mercury	Hg	–388	13.691	24.36	–	14.651	1.779
Molybdenum	Mo	2623	9.34	–	–	10.3	–
Neodymium	Nd	1021	6.688	5.3	–	21.57	0.79
			6.89[b]	7.6[b]	–	20.9	1.1
Nickel	Ni	1455	7.905	11.9	8.7–12.5	7.425	1.51
			7.850[a]	12.0[a]	–	7.477	1.53
Niobium	Nb	2477	7.83	–	–	11.9	–
Palladium	Pd	1554.9	10.49	12.3	–	10.14	1.17
Platinum	Pt	1768.4	18.91	28.8	–	10.32	1.52
Plutonium	Pu	640	16.65	14.1	14.1–15.2	14.66	0.847

continued

Table 17.4 (a) *(continued)*

Element		T_m °C	ρ_m 10^3 kg m^{-3}	$-\Lambda^{\dagger}$ 10^{-1} kg m^{-3} K^{-1}	Range‡	V_m 10^{-6} m^3 mol^{-1}	$\alpha_m^{\dagger\dagger}$ 10^{-4} K^{-1}
Potassium	K	63.4	0.827	2.4	2.2–2.5	47.3	2.9
Praseodymium	Pr	931	6.611	2.5	–	21.31	0.38
			6.50[b]	9.3[b]	–	21.7	1.4
Rhenium	Re	3186	18.8	–	–	9.90	–
Rhodium	Rh	1964	10.8	–	–	9.53	–
Rubidium	Rb	39.3	1.48	4.5	4.1–4.8	57.7	3.0
Ruthenium	Ru	2334	10.9	–	–	9.27	–
Samarium	Sm	1074	7.42[b]	5.2[b]	–	20.3	0.70
			7.058[b]	9.39[b]	–	21.30	1.33
Scandium	Sc	1541	2.68[b]	5.4[b]	–	16.8	2.0
Selenium	Se	220.5	4.00	11.7	11.7–12.3	19.7	2.93
Silicon	Si	1412	2.53	3.5	–	11.1	1.4
			2.560[a]	3.0[a]	–	10.97	1.2
			2.5785[d]	2[d]	–	10.892	0.78
Silver*	Ag	961.8	9.33	9.1	9.1–9.7	11.6	0.98
Sodium	Na	97.7	0.927	2.35	2.25–2.45	24.8	2.54
Strontium	Sr	777	2.37	2.6	–	37.0	1.1
Sulphur	S	119.0	1.819	8.00	–	17.63	4.40
Tantalum	Ta	3017	15.0	–	–	12.1	–
Tellurium	Te	450	5.80	7.3	5.2–7.3	22.0	1.3
Terbium	Tb	1356	8.05[b]	13.9[b]	–	19.7	1.73
			7.679[b]	4.84[b]	–	20.70	0.630
Thallium	Tl	304	11.35	13.0	11.7–14.4	18.01	1.15
Thorium	Th	1750	10.5	–	–	22.1	–
Tin	Sn	231.9	6.98	6.1	5.4–7.1	17.0	0.87
Titanium	Ti	1668	4.13	2.3	2.3–7.0	11.6	0.56
			4.140[a]	2.25[a]	–	11.57	0.543

Table 17.4 (a) *(continued)*

Element		T_m °C	ρ_m 10^3 kg m^{-3}	$-\Lambda^{\dagger}$ 10^{-1} kg m^{-3} K^{-1}	Range‡	V_m 10^{-6} m^3 mol^{-1}	$\alpha_m^{\dagger\dagger}$ 10^{-4} K^{-1}
Tungsten	W	3422	17.6	–	–	10.4	–
Uranium	U	1135	17.27	10.3	–	13.78	0.596
Vanadium	V	1910	5.36	3.2	–	9.50	0.60
Ytterbium	Yb	819	6.72[b]	32.5[b]		25.8	4.84
			6.245[b]	6.99[b]	–	27.71	1.12
Yttrium	Y	1522	4.18[b]	2.9[b]	–	21.3	0.69
				2.12[b]	–	–	0.507
Zinc	Zn	419.5	6.575	9.8	9.5–11.1	9.945	1.5
			6.756[a]	9.8[a]	–	9.679	1.5
Zirconium	Zr	1855	5.8	–	–	16	–

† $-\Lambda \equiv -d\rho / dT$.
‡ Experimental values for Λ (cf. Steinberg [6]).
†† $\alpha_m \equiv -\Lambda / \rho_m$.
Data, except for the those bearing the superscripts a, b, c, or d, are taken from Steinberg [6], Iida and Guthrie [4], and Gale and Tolemeier [1].
[a] Mills [7].
[b] Kononenko et al. [8].
[c] Iida et al. [9].
[d] Higuchi et al. [10].
* See Brillo et al. [11].

Table 17.4 (b) *Density data for high melting point liquid metals.*

Element		T_m °C	ρ_m 10^3 kg m^{-3}	$-\Lambda^{\dagger}$ 10^{-1} kgm^{-3}K^{-1}	V_m 10^{-6} m^3 mol^{-1}	$\alpha_m^{\ddagger}$ 10^{-4} K^{-1}	Temp.†† °C
Hafnium	Hf	2231	11.82	5.5	15.10	0.47	2027–2427
			12.0		14.9		1963
			11.1		16.1		2231
			11.97		14.91		2231
			11.5		15.5		2231
Iridium	Ir	2446	19.5	8.5	9.86	0.44	2027–2727
			20.0		9.61		2446
			19.39		9.913		2446

continued

Table 17.4 (b) *(continued)*

Element		T_m °C	ρ_m 10^3 kg m^{-3}	$-\Lambda^{\dagger}$ 10^{-1} kgm^{-3}K^{-1}	V_m 10^{-6} m^3 mol^{-1}	$\alpha_m^{\ddagger}$ 10^{-4} K^{-1}	Temp.†† °C
			19.23		9.996		2450
Molybdenum	Mo	2623	9.11	6.0	10.5	0.66	2177–2727
			9.35		10.3		2623
			9.10		10.5		2623
			9.33		10.3		2623
			9.10	8.0	10.5	0.88	2623 to
Niobium	Nb	2469	7.73	3.9	12.0	0.50	2027–2727
			7.83		11.9		2469
			7.57		12.3		2469
			7.6		12		2469
			7.68	5.4	12.1	0.70	2469 to
Osmium	Os	3033	19.1	11.6	9.96	0.607	2397–3107
			20.1		9.46		3033
			19.2		9.91		3033
Palladium	Pd	1555	10.66	7.7	9.983	0.72	1367–1602
			10.49	12.26	10.14	1.169	1555–1800
			10.7		9.95		
			10.379	11.69	10.253	1.126	1555–1700
			10.52		10.12		
Platinum	Pt	1768	19.22	9.6	10.15	0.50	1418–1943
			19.7		9.90		1768
			18.82		10.37		1800
			18.91	28.82	10.32	1.524	1768–1875
			18.81		10.37		
			19.77	24	9.867	1.2	1768–2200
			19.3	17	10.1	0.88	1768–4827

Table 17.4 (b) *(continued)*

Element		T_m °C	ρ_m 10^3 kg m^{-3}	$-\Lambda$† 10^{-1} kgm^{-3}K^{-1}	V_m 10^{-6} m^3 mol^{-1}	α_m‡ 10^{-4} K^{-1}	Temp.†† °C
			19.1	13	10.2	0.68	1822–4227
Rhenium	Re	3186	18.65	7.9	9.984	0.42	2410–3437
			18.7		9.96		3186
			18.0		10.3		3186
			18.9		9.85		3186
Rhodium	Rh	1963	10.82	7.6	9.511	0.70	1547–1977
			11.1		9.27		1963
			10.65		9.662		1963
			10.7	9.0	9.62	0.84	1963–2200
			10.7		9.62		1963
			12.2	5.0	8.43	0.41	1963–2200
Ruthenium	Ru	2334	10.75	5.6	9.402	0.52	1952–2502
			10.9		9.27		2334
Tantalum	Ta	3017	14.75	8.5	12.27	0.58	2377–3147
			15.0		12.1		3017
			14.43	13	12.54	0.90	
			14.6		12.4		3017
Tungsten	W	3422	16.43	10.8	11.19	0.657	2852–3434
			17.5		10.5		3420
			16.37	9.7	11.23	0.59	3420–7727
			16.26		11.31		3420
			16.2		11.35		3420–5067
			17.6		10.4		3420
Vanadium	V	1910	5.46	4.9	9.33	0.90	1567–1967
			5.55		9.18		1910
			5.73		8.89		1935

continued

Table 17.4 (b) *(continued)*

Element		T_m °C	ρ_m 10^3 kg m^{-3}	$-\Lambda^\dagger$ 10^{-1} kgm^{-3}K^{-1}	V_m 10^{-6} m^3 mol^{-1}	$\alpha_m{}^\ddagger$ 10^{-4} K^{-1}	Temp.†† °C
			5.36	3.2	9.50	0.60	1927–2197
			5.30		9.61		1910
			5.57		9.15		1902–6327
Yttrium	Y	1523	4.15	2.1	21.4	0.51	1287–1827
			4.15		21.4		
Zirconium	Zr	1855	6.21	2.7	14.7	0.43	1577–2477
			5.80		15.7		1855
			5.60		16.3		1855
			6.06		15.1		1835
			6.24	2.9	14.6	0.46	1427–2027
			5.50		16.6		1852

† $-\Lambda \equiv -d\rho/dT$.
‡ $\alpha_m \equiv -\Lambda/\rho_m$.
†† Temperature or temperature range.
Data are taken from Ishikawa et al. [12, 13].

17.5 Vapour Pressure Equations

Table 17.5 (a) *Vapour pressure equations:* $\log P(atm) = A + BT^{-1} + C\log T$ *(1 atm = 101,325 Pa).*

Element		A	B	C	Temperature range K
Aluminium	Al	5.911	−16211		m.p.–1800
Barium	Ba	4.007	−8163		m.p.–1200
Beryllium	Be	5.786	−15731		m.p.–1800
Cadmium	Cd	5.242	−5392		m.p.–650
Caesium	Cs	4.165	−3830		m.p.–550
Cerium	Ce	5.611	−21200		m.p.–2450
Cobalt	Co	6.488	−20578		m.p.–2150
Copper	Cu	5.849	−16415		m.p.–1850

Table 17.5 (a) *(continued)*

Element		*A*	*B*	*C*	Temperature range K
Curium	Cm	5.223	−18292		m.p.–2200
Erbium	Er	4.688	−14380		m.p.–1900
Gadolinium	Gd	5.557	−19389		m.p.–2250
Gallium	Ga	6.754	−13984	−0.3413	m.p.–1600
Gold	Au	5.832	−18024		m.p.–2050
Indium	In	5.374	−12276		m.p.–1500
Iron	Fe	6.347	−19574		m.p.–2100
Lanthanum	La	5.911	−21855		m.p.–2450
Lead	Pb	4.911	−9701		m.p.–1200
Lithium	Li	5.055	−8023		m.p.–1000
Lutetium	Lu	5.648	−20302		m.p.–2350
Mercury	Hg	5.116	−3190		298–400
Neodymium	Nd	4.912	−15824		m.p.–2000
Neptunium	Np	10.076	−23378	−1.3250	m.p.–2500
Nickel	Ni	6.666	−20765		m.p.–2150
Palladium	Pd	5.426	−17899		m.p.–2100
Platinum	Pt	6.386	−26856		m.p.–2500
Plutonium	Pu	3.666	−16658		m.p.–2450
Potassium	K	4.402	−4453		m.p.–600
Praseodymium	Pr	4.772	−17315		m.p.–2200
Protactinium	Pa	6.177	−32874		m.p.–2500
Rhodium	Rh	6.802	−26792		m.p.–2500
Rubidium	Rb	4.312	−4040		m.p.–550
Scandium	Sc	5.795	−17681		m.p.–2000
Silver	Ag	5.752	−13827		m.p.–1600
Sodium	Na	4.704	−5377		m.p.–700
Terbium	Tb	5.411	−18639		m.p.–2200

continued

Table 17.5 (a) *(continued)*

Element		*A*	*B*	*C*	Temperature range K
Thallium	Tl	5.259	–9037		m.p.–1100
Thorium	Th	–18.453	–24569	6.6473	m.p.–2500
Tin	Sn	5.262	–15332		m.p.–1850
Titanium	Ti	6.358	–22747		m.p.–2400
Uranium	U	20.735	–28776	–4.0962	m.p.–2500
Vanadium	V	6.929	–25011		m.p.–2500
Yttrium	Y	5.795	–20341		m.p.–2300
Zinc	Zn	5.378	–6286		m.p.–750
Zirconium	Zr	6.806	–30295		m.p.–2500

Data from Alcock et al. [14].

Table 17.5 (b) *Vapour pressure equations:* $\log P(mmHg) = A + BT^{-1} + C \log T$, *(1 mmHg = 133.322 Pa).*

Element		*A*	*B*	*C*	Temperature range K
Americium	Am	13.97	–13700	–1.0	1103–1453
Antimony (total pressure)	Sb_x	6.37	–6500		m.p.–b.p.
Bismuth	Bi	12.35	–10400	–1.26	m.p.–b.p.
	Bi_2	18.1	–10730	–3.2	m.p.–b.p.
Calcium	Ca	12.45	–8920	–1.39	m.p.–b.p.
Germanium	Ge	12.87	–18700	–1.16	m.p.–b.p.
Hafnium	Hf	9.20	–29830		m.p.–b.p.
Magnesium	Mg	12.79	–7550	–1.41	m.p.–b.p.
Manganese	Mn	19.24	–14520	–3.02	m.p.–b.p.
Phosphorus	P_4	7.84	–2740		m.p.–b.p.
Selenium (mainly Se_6)	Se_x	8.09	–4990		m.p.–b.p.

Table 17.5 (b) *(continued)*

Element		*A*	*B*	*C*	Temperature range K
Silicon	Si	10.78	–20900	–0.565	m.p.–b.p.
Strontium	Sr	12.63	–9000	–1.31	m.p.–b.p.
Sulphur (total pressure)	S_x	23.88	–4830	–5.0	m.p.–b.p.
Tellurrium	Te_2	22.29	–7830	–4.27	m.p.–b.p.

Data from Kubaschewski and Alcock [15].

17.6 Molar Heat Capacities at Constant Pressure

Table 17.6 (a) *Molar heat capacities at constant pressure.*

Element		Temperature or temperature range	Molar heat capacity, C_P J mol^{-1} K^{-1}			
		°C	A	B	C	D
Aluminium	Al	661	31.76	–	–	–
		660–900	–	29.1	–	–
		660.2–1000	–	–	31.8	–
Antimony	Sb	631	31.38	–	–	–
		630.4–1000	–	31.4	–	–
Barium	Ba	1027	40.58	–	–	–
		727	–	31.3	–	–
Beryllium	Be	1289	29.46	–	–	–
		1283	–	31.4	–	–
Bismuth	Bi	927	27.20	–	–	–
		271	–	30.5	–	–
		300	–	29.9	–	–
		400	–	30.82	–	–
		500	–	28.73	–	–

continued

Table 17.6 (a) *(continued)*

Element		Temperature or temperature range	Molar heat capacity, C_P J mol^{-1} K^{-1}			
		°C	A	B	C	D
		600	–	27.92	–	–
Boron	B	2077	30.54	31.5	–	–
Cadmium	Cd	321	29.71	–	–	–
		321–600	–	29.7	–	–
Caesium	Cs	427	30.94	–	–	–
		28.5	–	37	–	–
		100	–	35.2	–	–
		200	–	31.9	–	–
		400	–	28	–	–
		600	–	29	–	–
		800	–	33	–	–
Calcium	Ca	842	33.47	–	–	–
		865	–	31.1	–	–
Cerium	Ce	798	37.66	–	–	–
		804–1200	–	35	–	–
		804–2227	–	–	–	31.6
Chromium	Cr	1857	39.33	–	–	–
		1903	–	41	–	–
		1368–2122	–	–	–	50.71
Cobalt	Co	1495	40.50	–	–	48.55
		1493	–	35	–	–
		1495–1600	–	–	39.3	–
Copper	Cu	1085	32.84	–	–	–
		1083–1600	–	31.5	–	–
		1084.5–1500	–	–	31.5	–
		1115–1614	–	–	–	33.0

Table 17.6 (a) *(continued)*

Element		Temperature or temperature range	Molar heat capacity, C_P J mol^{-1} K^{-1}			
		°C	A	B	C	D
Dysprosium	Dy	1412	49.92	–	–	–
Erbium	Er	1522	38.70	–	–	–
Europium	Eu	823	38.07	–	–	–
Francium	Fr	18	–	31.7	–	–
		700	–	29.9	–	–
Gadolinium	Gd	1313	37.15	–	–	–
		1350	–	33.5	–	–
Gallium	Ga	30	26.78	–	–	–
		29.7–300	–	27.7	–	–
Germanium	Ge	937	27.61	–	–	–
		934–1000	–	29.3	–	–
Gold	Au	1227	30.96	–	–	–
		1063–1400	–	29.3	–	–
Hafnium	Hf	2231	33.47	–	–	–
Holmium	Ho	1472	43.93	–	–	–
		1500	–	33.5	–	–
Indium	In	156.5–600	–	29.7	–	–
Iridium	Ir	2447	41.84	–	–	–
Iron	Fe	1536	46.02	44.4	–	–
		1538–1800	–	–	46.0	–
		1531–1869	–	–	–	43.05
Lanthanum	La	918	34.31	–	–	–
Lead	Pb	1027	28.62	–	–	–
		327	–	31.5	–	–
		400	–	29.8	–	–
		500	–	28.4	–	–
		600	–	28.0	–	–

continued

Table 17.6 (a) *(continued)*

Element		Temperature or temperature range	Molar heat capacity, C_P J mol^{-1} K^{-1}			
		°C	A	B	C	D
Lithium	Li	180.4	–	30.33	–	–
		200	–	30.24	–	–
		400	–	29.26	–	–
		600	–	28.91	–	–
		800	–	28.79	–	–
		1000	–	28.78	–	–
		1600	–	30.3	–	–
Lutetium	Lu	1663	47.91	–	–	–
Magnesium	Mg	650	34.31	–	–	–
		650–1000	–	33.1	32.1	–
Manganese	Mn	1244	46.02	–	–	–
Mercury	Hg	–39.02	–	28.5	–	–
		0	–	28.5	–	–
		20	–	27.9	–	–
		100	–	27.5	–	–
		500	–	27.5	–	–
Molybdenum	Mo	2623	40.33	–	–	34.3
		2607	–	55	–	–
Neodymium	Nd	1016	48.79	–	–	–
		1024	–	33.5	–	–
		1173–1973	–	–	–	44.0
Nickel	Ni	1455	38.91	–	–	39.3
		1454	–	36.4	–	–
		1455–1600	–	–	43.1	–
Niobium	Nb	2472	41.78	–	–	–
		1758–2886	–	–	–	41.8

Table 17.6 (a) *(continued)*

Element		Temperature or temperature range	Molar heat capacity, C_P J mol^{-1} K^{-1}			
		°C	A	B	C	D
Palladium	Pd	1552	34.73	–	–	–
		1554.9	–	–	–	41.0
Phosphorus	P	44	26.32	–	–	–
Platinum	Pt	1769	34.69	–	–	–
		1770	–	34.7	–	–
		1929–2358	–	–	–	36.5
Potassium	K	63.4	–	32.1	–	–
		100	–	31.7	–	–
		200	–	30.9	–	–
		500	–	29.8	–	–
		1000	–	32.8	–	–
Praseodymium	Pr	918	42.97	–	–	–
		935	–	33.5	–	–
		1187–2016	–	–	–	41.57
Radium	Ra	960	–	30.7	–	–
Rhodium	Rh	1963	41.84	–	–	–
Rubidium	Rb	38.7	–	34.0	–	–
		100	–	32.7	–	–
		200	–	31.1	–	–
		500	–	29.7	–	–
		1000	–	32.3	–	–
Ruthenium	Ru	2250	41.84	–	–	–
Samarium	Sm	1073	50.21	–	–	–
		1072	–	33.5	–	–

continued

Table 17.6 (a) *(continued)*

Element		Temperature or temperature range	Molar heat capacity, C_P J mol^{-1} K^{-1}			
		°C	A	B	C	D
Scandium	Sc	1541	44.35	–	–	–
		1539	–	33.5	–	–
Selenium	Se	220	33.15	–	–	–
		217	–	35.1	–	–
Silicon	Si	1412	27.20	–	–	–
		1410–1600	–	29.2	–	–
		1414–1600	–	–	27.2	–
Silver	Ag	962	33.47	–	–	–
		960.6–1300	–	30.5	–	–
Sodium	Na	97	–	31.86	–	–
		100	–	31.84	–	–
		200	–	30.81	–	–
		400	–	29.38	–	–
		600	–	28.85	–	–
		800	–	29.20	–	–
		1000	–	30.25	–	–
		1200	–	32.30	–	–
Strontium	Sr	769	35.15	–	–	–
		770	–	31.0	–	–
Sulphur	S	115	35.19	–	–	–
		119	–	31.6	–	–
Tantalum	Ta	3020	41.84	–	–	–
		2380–3085	–	–	–	42.28
Tellurium	Te	450	37.66	–	–	–
		450–1000	–	37.6	–	–

Table 17.6 (a) *(continued)*

Element		Temperature or temperature range	Molar heat capacity, C_P J mol^{-1} K^{-1}			
		°C	A	B	C	D
Terbium	Tb	1360	46.44	–	–	–
Thallium	Tl	304	30.12	–	–	–
		303–500	–	30.5	–	–
Thorium	Th	1755	46.02	–	–	–
Thulium	Tm	1545	41.42	–	–	–
Tin	Sn	427	28.45	–	–	–
		232	–	29.7	–	–
		300	–	28.7	–	–
		400	–	28.6	–	–
		500	–	28	–	–
		1000	–	31	–	–
Titanium	Ti	1670	41.84	–	–	–
		1680	–	33.5	–	–
		1668–1800	–	–	46.2	–
		1696–2042	–	–	–	45.5
Tungsten	W	3420	54.02	–	–	–
		3422	–	–	–	53.0
Uranium	U	1135	48.66	–	–	48
		1133–1300	–	38.3	–	–
Vanadium	V	1910	47.28	–	–	–
		1912	–	39.7	–	–
		1932–2365	–	–	–	48.7
Ytterbium	Yb	819	36.82	–	–	–
Yttrium	Y	1522	39.79	–	–	–
		1530	–	33.5	–	–

continued

Table 17.6 (a) *(continued)*

Element		Temperature or temperature range	Molar heat capacity, C_P J mol^{-1} K^{-1}			
		°C	A	B	C	D
		1526–2087	–	–	–	39.8
Zinc	Zn	420	31.38	–	–	–
		419.4–800	–	31.5	–	–
		419.4–1000	–	–	31.3	–
Zirconium	Zr	1855	33.47	–	–	–
		1850	–	33.5	–	–
		1852	–	–	–	40.7
		2727	–	–	–	64.0

Sources of data: A, Kubaschewski et al. [16]; B, Gale and Tolemeier [1]; C, Mills [7]; D, Baykara [17].

Table 17.6 (b) *Molar heat capacities at constant pressure C_P for liquid metallic elements as a function of temperature T (K).*

Element		Melting point, T_m K	Molar heat capacity, C_P J mol^{-1} K^{-1}
Barium	Ba	1003	$22.84 + 7.82 \times 10^{-3}T + 126.78 \times 10^{5}T^{-2}$
Bismuth	Bi	545	$19.04 + 10.38 \times 10^{-3}T + 20.75 \times 10^{5}T^{-2} - 3.97 \times 10^{-6}T^{2}$
Caesium	Cs	302	$29.89 + 0.9 \times 10^{-3}T + 2.03 \times 10^{5}T^{-2}$
Gold	Au	1338	$50.33 - 12.69 \times 10^{-3}T$
Indium	In	430	$30.29 - 1.38 \times 10^{-3}T$
Lead	Pb	601	$32.49 - 3.10 \times 10^{-3}T$
Lithium	Li	454	$31.21 - 5.27 \times 10^{-3}T + 2.05 \times 10^{5}T^{-2} + 2.64 \times 10^{-6}T^{2}$
Mercury	Hg	298†	$30.38 - 11.46 \times 10^{-3}T + 10.15 \times 10^{-6}T^{2}$
Potassium	K	337	$37.18 - 19.12 \times 10^{-3}T + 12.30 \times 10^{-6}T^{2}$
Rubidium	Rb	313	$40.88 - 26.21 \times 10^{-3}T + 0.33 \times 10^{5}T^{-2} + 14.14 \times 10^{-6}T^{2}$
Sodium	Na	371	$37.51 - 19.22 \times 10^{-3}T + 10.64 \times 10^{-6}T^{2}$
Tin	Sn	505	$32.84 - 6.28 \times 10^{-3}T$

Data from Kubaschewski et al. [16].
† T_m = 234.3K (cf. Figure 4.10(a)).

17.7 Sound Velocity Data

Table 17.7 (a) *Sound velocity data (i.e. sound velocity U_m at the melting point T_m and the temperature coefficient † dU/dT of sound velocity).*

Element (metal)		T_m °C	U_m m s^{-1}	$-dU/dT$ m s^{-1} K^{-1}
Aluminium	Al	660.3	4680	0.474
Barium	Ba	727	1331	0.177
Beryllium	Be	1287	9104	0.72
Cadmium	Cd	321.1	2237	0.390
Calcium	Ca	842	2978	0.495
Cobalt	Co	1495	4031	0.461
Copper	Cu	1084.6	3440	0.451
Gallium	Ga	29.8	2873	0.288
Gold	Au	1064.2	2568	0.571
Indium	In	156.6	2320	0.289
Iron	Fe	1538	4200	0.499
Lanthanum	La	918	2022	0.078
Lead	Pb	327.5	1821	0.317
Lithium	Li	180.5	4554	0.602
Magnesium	Mg	650	4065	0.590
Manganese	Mn	1246	3381	0.593
Mercury	Hg	−38.8	1511	0.483
Molybdenum	Mo	2623	4502	0.467
Nickel	Ni	1455	4047	0.387
Platinum	Pt	1768.4	3053	0.241
Praseodymium	Pr	931	1925	0.084
Silver	Ag	961.8	2790	0.394
Strontium	Sr	777	1902	0.310
Tantalum	Ta	3017	3303	0.256

continued

Table 17.7 (a) *(continued)*

Element (metal)		T_m °C	U_m m s^{-1}	$-dU/dT$ m s^{-1} K^{-1}
Thallium	Tl	304	1650	0.239
Tin	Sn	231.9	2464	0.277
Tungsten	W	3422	3279	0.467
Ytterbium	Yb	819	1274	0.198
Zinc	Zn	419.5	2850	0.337

† Sound velocity, $U = a - bT$, where a and b are constants; $dU/dT = -b$, Data from Blairs [18,19].

Table 17.7 (b) *Sound velocity data (i.e. sound velocity U_m at the melting point T_m and U at temperature T in kelvin).*

Metallic element		T_m °C	D_m m s^{-1}	U m s^{-1}
Antimony†	Sb	630.6	1910	$1330.6 + 1.0471T - 4.4946 \times 10^{-4}T^2$
Bismuth	Bi	271.4	1640	$1720 - 9.199 \times 10^{-2}T - 4.628 \times 10^{-5}T^2$
Caesium	Cs	28.4	983	$1027.5 - 0.214T - 7.123 \times 10^{-5}T^2$
Cerium†	Ce	798	1693	$1503 + 0.177T$
Germanium†	Ge	937	2693	$-9192.5 + 25.751T - 1.8383 \times 10^{-2}T^2 + 4.3209 \times 10^{-6}T^3$
Plutonium†	Pu	640	1195	$1119 + 0.083T$
Potassium	K	63.4	1876	$2034.4 - 0.430T - 7.700 \times 10^{-5}T^2$
Rubidium	Rb	39.3	1251	$1324.5 - 0.2313T - 1.057 \times 10^{-4}T^2$
Selenium	Se	220.5	1100	$1599.8 - 1.340T + 2.775 \times 10^{-4}T^2$
Silicon†	Si	1412	3920	$-2131 + 6.238T - 1.574 \times 10^{-3}T^2$
Sodium	Na	97.7	2526	$2660.7 - 0.37667T - 9.0366 \times 10^{-5}T^2$
Tellurium†	Te	450	889	$-402.6 + 2.7788T - 1.2512 \times 10^{-3}T^2$

† The velocity of sound increases with increasing temperature, passes through a maximum, and then decreases at higher temperatures. Data from Blairs [19].

Recently, Hayashi et al. [20] measured the velocity of sound in Group IVA metallic elements. Their experimental results are shown in Table 17.7(c). As can be seen, Hayashi et al.'s experimental data are in good agreement with the above data given by Blairs

Table 17.7 (c) *Sound velocity data for Group IVA metallic elements (T in kelvin).*

Metallic elements		T_m °C	U_m m s^{-1}	U m s^{-1}
Germanium	Ge	937	2690	$891 + 2.88T - 1.15 \times 10^{-3} T^2$
Lead	Pb	327.5	1820	$1990 - 0.286T$
Silicon	Si	1412	3900	$-970 + 4.86T - 1.17 \times 10^{-3} T^2$
Tin	Sn	231.9	2490	$2630 - 0.269T$

17.8 Surface Tension Data

Table 17.8 *Surface tension data (i.e. surface tension*† γ_m *at the melting point Tm and the temperature coefficient* $d\gamma / dT$ *of surface tension).*

Element		T_m °C	γ_m mN m^{-1}	$-d\gamma / dT$ mN m^{-1} K^{-1}
Aluminium	Al	660.3	1050	0.25
Antimony	Sb	630.6	371	0.045
Barium	Ba	727	273	0.072
Beryllium	Be	1287	1100 (at 1500 °C)	–
Bismuth	Bi	271.4	382	0.08
Cadmium	Cd	321.1	637	0.15
Caesium	Cs	28.4	69.0	0.047
Calcium	Ca	842	363	0.10
Cerium	Ce	798	794	0.074
Chromium	Cr	1907	1710	0.54
Cobalt	Co	1495	1900	0.35
Copper	Cu	1084.6	1320	0.28
Dysprosium	Dy	1412	648	0.13
Erbium	Er	1529	637	0.12
Europium	Eu	822	264	0.05
Francium[a]	Fr	27	62	0.044

continued

Table 17.8 *(continued)*

Element		T_m °C	γ_m mN m^{-1}	$-d\gamma/dT$ mN m^{-1} K^{-1}
Gadolinium	Gd	1313	664	0.058
Gallium	Ga	29.8	724	0.07
Germanium	Ge	937	607	0.14
Gold	Au	1064.2	1150	0.15
Hafnium	Hf	2233	1614	0.100
Holmium	Ho	1474	650	0.123
Indium	In	156.6	560	0.090
Iridium	Ir	2446	2250	0.25
Iron	Fe	1538	1880	0.41
Lanthanum	La	918	728	0.10
Lead	Pb	327.5	457	0.11
Lithium	Li	180.5	399	0.15
Lutetium	Lu	1663	940	0.073
Magnesium	Mg	650	577	0.26
Manganese	Mn	1246	1152	0.35
Mercury	Hg	−38.8	489	0.23
Molybdenum	Mo	2623	2080	–
Neodymium	Nd	1021	687	0.088
Nickel	Ni	1455	1795	0.33
Niobium	Nb	2477	1937	0.20
Osmium	Os	3033	2400	–
Palladium	Pd	1554.9	1482	0.279
Platinum	Pt	1768.4	1746	0.307
Plutonium	Pu	640	550 (at 640 °C)	–
Potassium	K	63.4	110	0.07
Praseodymium	Pr	931	716	0.08
Rhenium	Re	3186	2610 (±100)	–

Table 17.8 *(continued)*

Element		T_m °C	γ_m mN m^{-1}	$-d\gamma/dT$ mN m^{-1} K^{-1}
Rhodium	Rh	1964	1940	0.30
Rubidium	Rb	39.3	89.6	0.056
Ruthenium	Ru	2334	2260	0.24
Samarium	Sm	1074	430	0.072
Scandium	Sc	1541	939	0.124
Selenium	Se	220.5	103	0.15
Silicon	Si	1412	850 (± 20)	0.25
Silver	Ag	961.8	926	0.22
Sodium	Na	97.7	197	0.09
Strontium	Sr	777	296	0.08
Tantalum	Ta	3017	2050	–
Tellurium	Te	450	239	0.104
Terbium	Tb	1356	669	0.056
Thallium	Tl	304	459	0.106
Thorium	Th	1750	1006 (at 1690 °C)	–
Tin	Sn	231.9	555	0.07
Titanium	Ti	1668	1670	0.16
Tungsten	W	3422	2310	–
Uranium	U	1135	1552	0.27
Vanadium	V	1910	1860	–
Ytterbium	Yb	819	320	0.102
Yttrium	Y	1522	872	0.086
Zinc	Zn	419.5	789	0.21
Zirconium	Zr	1855	1500	0.11

† Surface tension, $\gamma = \gamma_m + \frac{d\gamma}{dT}(T - T_m)$.
Data, except for those bearing the superscript a, are taken from Keene [21] and Mills and Su [22].
[a] From Gale and Tolmeier [1].
See also Ishikawa and Paradis [13].

17.9 Viscosity Data

Table 17.9 (a) *Viscosity data.*

Metallic element		T_m† K	$\mu_m^{\ddagger}$ mPa s	A†† mPa s	B†† kJ mol^{-1}	C_A‡‡ 10^{-7} kg$^{1/2}$ m s^{-1} K$^{-1/2}$ mol$^{-1/6}$
Aluminium	Al	933.5	1.38	0.2565	13.08	1.39
Antimony	Sb	903.8	1.44	0.3042	11.67	0.970
Bismuth	Bi	544.6	1.59	0.5035	5.192	1.13
Cadmium	Cd	594.2	2.50	0.6438	6.698	1.78
Caesium	Cs	301.6	0.686	0.1014	4.79	1.88
Calcium	Ca	1115	1.10	0.0579	27.51	1.57
Cerium	Ce	1071	3.25	0.6748	13.97	2.02
Cobalt	Co	1768	5.4	0.2042	48.13	2.04
Copper	Cu	1357.8	4.38	0.5287	23.85	1.88
Francium	Fr	300	0.765	–	–	1.95
Gallium	Ga	302.9	1.95	0.6209	2.880	2.15
Germanium	Ge	1210	0.73	~0.23	11.5	0.44
Gold	Au	1337.3	5.37	1.099	17.62	1.67
Indium	In	429.7	1.79	0.4129	5.244	1.64
Iron	Fe	1811	5.6	0.2388	47.44	2.22
Lanthanum	La	1191	2.66	0.2092	25.22	1.69
Lead	Pb	600.6	2.59	0.5078	8.142	1.68
Lithium	Li	453.7	0.602	0.1386	5.52	1.91
Magnesium	Mg	923	1.25	0.025	30.5	1.63
Manganese	Mn	1519	5	0.12–1.02	20–46.5	2.46
Mercury	Hg	234.3	2.048	0.5353	2.614	1.79
Nickel	Ni	1728	4.7	0.3135	38.85	1.78
Palladium	Pd	1828.1	4.22 (1833 K)	–	–	1.42
Plutonium	Pu	913	5.95	1.089	12.88	2.39
Potassium	K	336.5	0.537	0.0999	4.69	1.94

Table 17.9 (a) *(continued)*

Metallic element		T_m† K	$\mu_m^{\ddagger}$ mPa s	A†† mPa s	B†† kJ mol^{-1}	C_A‡‡ 10^{-7} kg$^{1/2}$ m s^{-1} K$^{-1/2}$ mol$^{-1/6}$
Praseodymium	Pr	1204	2.85	0.9359	11.18	1.68
Rubidium	Rb	312.5	0.674	0.0927	5.13	1.95
Selenium	Se	493.7	24.8	~ 0.447	16.5	29.0
Silicon	Si	1685	0.58/0.8/0.94	0.06–0.12	27–36	0.42/0.6/0.68
Silver	Ag	1234.9	3.88	0.453	22.2	1.72
Sodium	Na	370.9	0.695	0.0916	6.25	2.03
Tellurium	Te	723	1.8–2.1	0.699	5.3	1.5–1.7
Thallium	Tl	577	2.59	0.5242	7.638	1.64
Tin	Sn	505.1	1.87	0.4670	5.833	1.60
Titanium [a]	Ti	1941	4.4	0.033	76.6	2.34
Uranium	U	1408	6.57	0.4849	30.45	2.06
Ytterbium	Yb	1092	2.67	0.1974	23.77	1.70
Yttrium [a]	Y	1796	3.6	0.00287	106.5	2.19
Zinc	Zn	692.7	3.50	0.5266	10.91	2.41

† Melting point
‡ Melting point viscosity
†† A and B are constants appearing in the following relation:
$\mu = A\exp\left(\frac{B}{RT}\right)$, (in mPa s)
where μ is the viscosity, R is the molar gas constant, and T is the absolute temperature (K).
‡‡ Andrade coefficient
$C_A = \mu_m \frac{V_m^{2/3}}{(MT_m)^{1/2}}$
where V_m is the molar volume at T_m and M is the molar mass.
Data, except for those bearing the superscript a, are taken from Iida et al. [23].
[a] Data from Ishikawa and Paradis [13].

Table 17.9 (b) *Viscosity data for high melting point liquid metals.*

Metal		T_m K	μ_m mPa s	A mPa s	B kJ mol^{-1}	C_A† 10^{-7} kg$^{1/2}$ m s^{-1} K$^{-1/2}$ mol$^{-1/6}$
Chromium [a]	Cr	2180	5.7	1.7×10^{-4}	~ 185	2.19
Hafnium	Hf	2504	5.2	0.50	48.7	1.57
			5.0	–	–	1.51
Iridium	Ir	2719	7.0	1.85	30.0	1.38
Molybdenum	Mo	2896	5.6	0.27	73	1.59
Niobium	Nb	2742	4.5	0.55	48.9	1.47
Osmium	Os	3306	4.2	0.00167	220	0.75
Platinum	Pt	2041	4.82	0.25	49.9	1.14
			6.74	1.53	25.263	1.60
Rhenium	Re	3459	7.9	0.08	133	1.44
Rhodium	Rh	2236	2.9	0.09	64.3	0.86
			5	–	–	1.48
Ruthenium	Ru	2607	6.1	0.60	49.8	1.66
Tantalum	Ta	3290	8.6	0.004	213	1.86
Tungsten	W	3695	6.9	0.11	128.	1.26
Vanadium [b]	V	2183	2.4(2163 *K*)	–	–	1.02
Zirconium	Zr	2128	4.7	0.76	31.8	2.14
			4.83			2.20
			3.5			1.60
			5.45			2.48

Data, except for those bearing the superscripts †, a or b, are from Ishikawa and Paradis [13].

[a] Batterzzati and Greer [24].

[b] Agaev et al. [25].

17.10 Self-Diffusivity Data

Table 17.10 *Self-diffusivity data.*

Metallic element		T_m K	D_m† 10^{-9} m² s⁻¹	D_0† 10^{-9} m² s⁻¹	H_D† kJ mol⁻¹
Antimony	Sb	903.8	5.18	5.46	17.70
Bismuth	Bi	544.6	3.78	3.83	10.49
			0.8002		–
Cadmium	Cd	594.2	1.78	–	–
Caesium	Cs	301.6	2.69	–	–
Copper	Cu	1357.8	4.00	14.6	40.6
Gallium	Ga	302.9	1.71	1.1	4.69
			1.60	3.45	7.74
			1.57	$D = 6.01 \times 10^{-13} T^{3/2} - 1.6 \times 10^{-9}$	
Germanium	Ge	1210	14.2	22	27.6
Indium	In	429.7	1.66	2.89	10.2
			1.67	3.34	10.7
Lead	Pb	600.6	2.21	9.15	18.6
			1.75	2.37	13.0
Lithium	Li	453.7	6.18	14.1	11.8
			7.34	9.4	9.62
			5.76	$D = \{5.76 + 0.036(T - T_m)\} \times 10^{-9}$	
			6.80	$D = \{6.80 + 0.036(T - T_m)\} \times 10^{-9}$	
			5.98	14.4	12.0
Mercury	Hg	234.3	1.08	1.3	4.85
			0.97	0.85	4.23
			1.11	$\log D = 1.854 \log T - 13.349$	
			0.93	1.1	4.81
			1.08	$D = 4.34 \times 10^{-13} T^{3/2} - 4.81 \times 10^{-10}$	
Potassium	K	336.5	3.71	17	10.7

continued

Table 17.10 *(continued)*

Metallic element		T_m K	D_m† 10^{-9} m^2 s^{-1}	D_0† 10^{-9} m^2 s^{-1}	H_D† kJ mol^{-1}
			3.71	7.6	8.45
			3.61	$D = 5.344 \times 10^{-14} T^2 - 2.443 \times 10^{-9}$	
Rubidium	Rb	312.5	2.63	5.7	7.99
			2.73	6.6	8.28
Silicon	Si	1685	40±5	–	–
Silver	Ag	1234.9	2.56	7.10	34.1
			2.57	5.8	32.0
Sodium	Na	370.9	4.03	11	10.2
			3.85	9.2	9.79
			4.23	8.6	9.29
Tellurium	Te	723	2.72	12.9	23.2
Thallium	Tl	577	2.009	–	–
Tin	Sn	505.1	2.31	3.02	10.8
			2.07	$D = \{0.0172(T - 504.9) + 2.07\} \times 10^{-9}$	
Zinc	Zn	692.7	2.03	8.2	21.3
			2.06	12	23.4

† D_m is the self–diffusivity at the melting point T_m and D_0 and H_D are Arrhenius parameters appearing in the following equation: $D = D_0 \exp\left(-\frac{H_D}{RT}\right)$
where D is the self-diffusivity at temperature T (in K) and R is the molar gas constant.
Data, except for one datum point of silicon, $(D_m)_{si}$, at the melting point, are taken from Iida et al. [26]. $(D_m)_{si}$ from Sanders and Aziz [27].

17.11 Electrical Resistivity and Thermal Conductivity Data

Table 17.11 *Electrical resistivity and thermal conductivity data.*

Element		Temperature	$\rho_{e,l}/\mu\Omega$ m		λ†
		°C	G.T.†	K.M.‡	W m^{-1} K^{-1}
Aluminium	Al	660	0.2425	0.200	94.05
		700	0.2483	0.204	95.37
		800	0.2630	0.213	98.71
		900	0.2777	0.222	102.05
		1000	0.2924	–	105.35
Antimony	Sb	630.4	1.135	1.135	21.8
		700	1.154	1.154	21.3
		800	1.181	1.181	20.9
		900	–	1.208	–
		1000	1.235	1.235	–
Arsenic	As	817	2.10	–	–
Barium	Ba	727	1.33	–	–
Beryllium	Be	1283	0.45	–	–
Bismuth	Bi	271	1.290	1.281	17.1
		300	–	1.319	15.5
		400	–	1.376	15.5
		500	–	–	15.5
		600	–	1.490	15.5
		700	–	–	15.5
		800	–	1.604	–
Boron	B	2077	2.10	–	–
Cadmium	Cd	321	0.337	0.337	42
		350	–	0.337	–
		400	0.3430	0.337	47

continued

Table 17.11 *(continued)*

Element		Temperature	$\rho_{e,l}/\mu\Omega$ m		$\lambda^{\dagger}$
		°C	G.T.†	K.M.‡	W m^{-1} K^{-1}
		500	0.3510	0.341	54
		600	0.3607	0.348	61
Caesium	Cs	28.5	0.370	–	19.7
		29.7	–	0.366	–
		37	–	0.370	–
		80	–	0.385	–
		100	0.450	–	20.2
		200	0.565	–	20.8
		400	0.810	–	20.2
		600	1.125	–	18.3
		800	1.570	–	16.1
		1600	–	–	4.0
Calcium	Ca	865	0.250	–	–
Cerium	Ce	759	–	1.28	–
		800	–	1.28	–
		804	1.268	–	–
		900	–	1.29	–
		1000	1.294	–	–
		1200	1.310	–	–
Chromium	Cr	1903	0.316	–	–
Cobalt	Co	1493	1.02	–	–
Copper	Cu	1083	0.200	0.211	165.6
		1100	0.202	0.212	166.1
		1150	–	0.216	–
		1200	0.212	0.221	170.1
		1400	0.233	–	176.3

Table 17.11 *(continued)*

Element		Temperature	$\rho_{e,l}/\mu\Omega$ m		$\lambda^{\dagger}$
		°C	G.T.[†]	K.M.[‡]	W m^{-1} K^{-1}
		1600	0.253	–	180.4
Francium	Fr	18	0.87	–	–
Gadolinium	Gd	1310	–	1.95	–
		1350	0.278	–	–
		1400	–	1.96	–
		1600	–	1.96	–
Gallium	Ga	29.7	0.26	0.258	25.5
		50	–	0.262	–
		100	0.27	0.272	30.0
		200	0.28	0.292	35.0
		300	0.30	0.311	39.2
		500	–	0.331	–
Germanium	Ge	934	0.672	–	–
		1000	0.727	–	–
Gold	Au	1063	0.3125	–	104.44
		1100	0.3180	–	105.44
		1200	0.3315	–	108.15
		1300	0.3481	–	110.84
		1400	0.3631	–	113.53
Hafnium	Hf	2227	2.18	–	–
Holmium	Ho	1500	1.93	–	–
Indium	In	156.5	0.3230	0.324	42
		200	0.3339	0.334	–
		300	–	0.359	–
		400	0.4361	0.384	–

continued

Table 17.11 *(continued)*

Element		Temperature	$\rho_{e,l}/\mu\Omega$ m		$\lambda^{\dagger}$
		°C	G.T.[†]	K.M.[‡]	W m^{-1} K^{-1}
		600	0.5131	–	–
Iron	Fe	1536	1.386	–	–
Lanthanum	La	930	1.38	–	21.0
		1000	1.43	–	–
		1100	1.50	–	–
		1200	1.56	–	–
Lead	Pb	327	0.9485	0.950	15.4
		350	–	0.962	–
		400	0.9863	0.987	16.6
		450	–	1.012	–
		500	1.0344	–	18.2
		520	–	1.036	–
		550	–	1.059	–
		600	1.0825	1.083	19.9
		650	–	1.106	–
		700	–	1.129	–
		750	–	1.152	–
		800	1.169	1.174	–
		1000	1.263	–	–
Lithium	Li	180.4	0.240	0.255	46.4
		200	–	0.265	47.2
		250	–	0.285	–
		300	–	0.300	–
		400	–	–	53.8
		600	–	–	57.5
		800	–	–	58.6

Table 17.11 *(continued)*

Element		Temperature	$\rho_{e,l}/\mu\Omega$ m		$\lambda^{\dagger}$
		°C	G.T.[†]	K.M.[‡]	W m^{-1} K^{-1}
		1000	–	–	58.4
		1600	–	–	52.0
Magnesium	Mg	650	0.274	–	78
		700	0.277	–	81
		800	0.282	–	88
		1000	–	–	100
Mercury	Hg	−39.02	0.905	–	6.78
		0	0.940	–	7.61
		20	0.957	–	8.03
		50	–	0.984	–
		100	1.033	1.032	9.47
		200	–	1.142	–
		300	–	1.275	–
		350	–	1.355	–
		500	1.600	–	12.67
		1000	3.77	–	8.86
		1460	~1000	–	~ 0.0004
Molybdenum	Mo	2607	0.605	–	–
Neodymium	Nd	1024	1.26	–	–
Nickel	Ni	1454	0.850	–	–
Niobium	Nb	2468	1.05	–	–
Phosphorus	P	44	2.70	–	–
Platinum	Pt	1770	0.73	–	–
Plutonium	Pu	640	1.33	–	–
Polonium	Po	254	3.98	–	–
Potassium	K	63.4	0.1365	0.132	53.0

continued

Table 17.11 *(continued)*

Element		Temperature	$\rho_{e,l}/\mu\Omega$ m		$\lambda^{\dagger}$
		°C	G.T.[†]	K.M.[‡]	W m^{-1} K^{-1}
		100	0.154	0.154	51.7
		200	0.215	0.218	47.7
		300	–	0.282	–
		500	0.444	–	37.8
		1000	0.110	–	24.4
		1500	–	–	15.5
Praseodymium	Pr	935	1.38	–	–
Radium	Ra	960	1.71	–	–
Rhenium	Re	3158	1.45	–	–
Rubidium	Rb	38.7	0.2283	–	33.4
		50	–	0.232	–
		100	0.2730	0.275	33.4
		200	0.3665	–	31.6
		500	0.6890	–	26.1
		1000	1.71	–	17.0
		1500	5.32	–	8.0
Ruthenium	Ru	2427	0.84	–	–
Samarium	Sm	1072	1.90	–	–
Scandium	Sc	1539	1.31	–	–
Selenium	Se	217	$\sim 10^6$	–	0.3
Silicon	Si	1410	0.75	–	–
		1500	0.82	–	–
		1600	0.86	–	–
Silver	Ag	960.6	0.1725	–	174.8
		1000	0.1760	–	176.5
		1100	0.1845	–	180.8
		1200	0.1935	–	185.1

Table 17.11 *(continued)*

Element		Temperature	$\rho_{e,l}/\mu\Omega$ m		$\lambda^{\dagger}$
		°C	G.T.[†]	K.M.[‡]	W m^{-1} K^{-1}
		1300	0.2023	–	189.3
		1400	0.2111	–	193.5
Sodium	Na	97	0.0964	–	89.7
		100	0.099	–	89.6
		200	0.134	–	82.5
		400	0.224	–	71.6
		600	0.326	–	62.4
		800	0.469	–	53.7
		1000	–	–	45.8
		1200	–	–	38.8
Strontium	Sr	770	0.58	–	–
Sulphur	S	119	$>10^{10}$	–	–
Tantalum	Ta	2996	1.18	–	–
Tellurium	Te	450	5.50	–	2.5
		500	4.80	–	3.0
		600	4.30	–	4.1
		800	3.9	–	6.2
		1000	3.8	–	–
Terbium	Tb	1360	–	1.90	–
		1365	2.44	–	–
		1400	–	1.93	–
		1600	–	1.96	–
Thallium	Tl	303	0.731	–	24.6
		400	0.759	–	–
		500	0.788	–	–
Thulium	Tm	1600	1.88	–	–

continued

Table 17.11 *(continued)*

Element		Temperature	$\rho_{e,l}/\mu\Omega$ m		$\lambda^{\dagger}$
		°C	G.T.†	K.M.‡	W m^{-1} K^{-1}
Tin	Sn	232	0.4720	0.473	30.0
		250	–	0.477	–
		300	0.4906	0.490	31.4
		350	–	0.504	–
		400	0.5171	0.518	33.4
		450	–	0.531	–
		500	0.5435	0.544	35.4
		530	–	0.553	–
		1000	0.670	–	–
Titanium	Ti	1685	1.72	–	–
Tungsten	W	3377	1.27	–	–
Uranium	U	1133	0.636	–	–
		1200	0.653	–	–
		1300	0.678	–	–
Vanadium	V	1912	0.71	–	–
Ytterbium	Yb	824	1.64	–	–
Yttrium	Y	1530	1.04	–	–
Zinc	Zn	419.4	0.374	–	49.5
		500	0.368	–	54.1
		600	0.363	–	59.9
		800	0.367	–	60.7
Zirconium	Zr	1850	1.53	–	–

† Data from Gale and Tolemeier [1].
‡ Data from Kozuka and Morinaga [28].

REFERENCES

1. W.F. Gale and T.C. Tolemeier (eds.), *Smithells Metals Reference Book*, 8th ed., Elsevier Butterworth-Heinemann, Oxford, 2004, 14–1.

2. W. Benenson, J.W. Harris, H. Stocker, and H. Lutz (eds.), *Handbook of Physics*, Springer-Verlag, New York, 2002.
3. S. Nagakura, H. Inokuchi, H. Ezawa, H. Iwamura, F. Sato, and R. Kubo (eds.), *Iwanami Dictionary of Physical Sciences (Iwanami Rikagaku Jiten)*, 5th ed., Iwanami Shoten Publishers, Tokyo, 1998.
4. T. Iida and R.I.L. Guthrie, *The Physical Properties of Liquid Metals*, Clarendon Press, Oxford, 1988.
5. The Japan Institute of Metals (ed.), *Metals Data Book*, 4th ed., Maruzen Company, Tokyo, 2004.
6. D.J. Steinberg, *Met. Trans.*, 5 (1974), 1341.
7. K.C. Mills, *Recommended Values of Thermophysical Properties for Selected Commercial Alloys*, Woodhead Publishing, Cambridge, 2002.
8. V.I. Kononenko, A.L. Sukhman, S.L. Gruverman, and V.V. Torokin, *Phys. Stat. Sol. (a)*, **84** (1984), 423.
9. T. Iida, Y. Kita, Y. Kikuya, T. Kirihara, and Z. Morita, *J. Non-Cryst. Solids*, **117–118** (1990), 567.
10. K. Higuchi, K. Kimura, A. Mizuno, M. Watanabe, Y. Katayama, and K. Kuribayashi, *Meas. Sci., Technol.*, **16** (2005), 381.
11. J. Brillo, I. Egry, and I. Ho, *Int. J. Thermophys.*, **27** (2006), 494.
12. T. Ishikawa, P.-F. Paradis, T. Itami, and S. Yoda, *Meas. Sci. Technol.*, **16** (2005), 443.
13. T. Ishikawa and P.-F. Paradis, in *High-Temperature Measurements of Materials*, H. Fukuyama and Y. Waseda (eds.), Springer, Berlin, 2009, p. 173.
14. C.B. Alcock, V.P. Itkin, and M.K. Horrigan, *Can. Met. Q.*, **23** (1984), 309.
15. O. Kubaschewski and C.B. Alcock, *Metallurgical Thermochemistry*, 5th ed., Pergamon Press, Oxford, 1979.
16. O. Kubaschewski, C.B. Alcock, and P.J. Spencer, *Materials Thermochemistry*, 6th ed., Pergamon Press, Oxford, 1993.
17. T. Baykara, R.H. Hauge, N. Norem, P. Lee, and J.L. Margrave, *A Review of Containerless Thermophysical Property Measurements of Liquid Metals and Alloys*, II Ciocco Workshop Proc., 1993.
18. S. Blairs, *Phys. Chem. Liq.*, **45** (2007), 399.
19. S. Blairs, *Int. Mater. Rev.*, **52** (2007), 321.
20. M. Hayashi, H. Yamada, N. Nabeshima, and K. Nagata, *Int. J. Thermophys.*, **28** (2007), 83.
21. B.J. Keene, *Int. Mater. Rev.*, 38 (1993), 157.
22. K.C. Mills and Y.C. Su, *Int. Mater. Rev.*, **51** (2006), 329.
23. T. Iida, R. Guthrie, M. Isac, and N. Tripathi, *Metall. Mater. Trans B*, 37 (2006), 403.
24. L. Battezzati and A.L. Greer, *Acta Metall.*, **37** (1989), 1791.
25. A.D. Agaev, V.I. Kostikov, and V.N. Bobkovskii, *Izv. A.N. SSSR, Metally*, **3** (1980), 43.
26. T. Iida, R. Guthrie, and N. Tripathi, *Metall. Mater. Trans. B*, 37 (2006), 559.
27. P.G. Sanders and M.J. Aziz, *J. Appl. Phys.*, **86** (1999), 4258.
28. Z. Kozuka and K. Morinaga, in *Handbook of Physico-Chemical Properties at High Temperatures*, edited by Y. Kawai and Y. Shiraishi, Iron and Steel Institute of Japan, Tokyo, 1988, p. 205.

Glossary and/or Supplementary Explanations

Absolute temperature or thermodynamic temperature (symbol T) a scale of temperature based on absolute zero, i.e. $T = 0$ K = –273.15°C. The unit in which absolute temperature, or thermodynamic temperature, is expressed as the kelvin. Temperature is a measure of the kinetic energy of the molecules, atoms, or ions of which a body or substance is composed.

In quantum theory, absolute zero is interpreted as the temperature at which all particles (e.g. atoms, molecules) are in the lowest-energy states. At absolute zero, they no longer have excitation energy, and all lattice oscillations in a solid are frozen; nevertheless, at $T = 0$, the kinetic energy is not zero (because of Heisenberg's uncertainty relation), since there is still molecular energy (called the zero-point energy) and the molecules carry out quantum-mechanical zero-point oscillations.

According to Charles' law, the volume V (= kT, where k is a constant) of (an ideal) gas would be zero at 0 K, but this has never been accomplished, because all known gases become liquid before reaching that temperature.

Actinoid contraction a smooth decrease in atomic or ionic radius with increasing proton number, or atomic number, found in the actinoids.

When we move across the actinoid series (cf. Table 1.7 or Table 10.4), the number of protons increases, along with the number of f-electrons; the actinoid contraction is due to pulling of these inner f-electrons towards the centre of the atoms by the increased nuclear charge (see Lanthanoid contraction).

Actinoids a series of elements in the periodic table from actinium ($_{89}$Ac) to lawrencium ($_{103}$Lr) inclusive. Actinoids is a general term for these 15 elements (or metals). The actinoids all have two outer s-electrons, and they are classified together by the fact that increasing proton number corresponds to the filling of the 5f level. In fact, because the 5f and 6d levels are close in energy, the filling of the 5f orbitals is not smooth (cf. Table 1.5). The first four members ($_{89}$Ac to $_{92}$U) exist naturally on the Earth. All are radioactive and this makes investigation difficult because of self-heating, short life-time, etc. The lighter members of the series (up to americium) have f-electrons that can participate in bonding, unlike the lanthanoids. Consequently, these elements resemble the transition metals in forming coordination complexes and displaying variable valence. Actinium itself does not have a 5f electron ($6d^1 7s^2$), but it is usually classified with actinoids because of its chemical similarities.

Amount of substance a measure of the number of particles in a quantity of equal particles (e.g. atoms, molecules, ions), independent of their mass. In other words, the amount of substance of an element is proportional to the number of particles (e.g. atoms) present. The constant of proportionality is the same for all substances and is the reciprocal of the Avogadro constant, namely, amount of substance = particle number/ Avogadro constant. The SI unit of the amount of substance is the mole; 1 mol is the quantity of substance that contains just as many particles as 0.012 kg ^{12}C.

Anharmonic motion the motion of a body or substance subjected to a restoring force that is not directly proportional to the displacement from a fixed point in the line of motion. Anharmonic motions of atoms (molecules, ions) play an important role to accurately understand the thermophysical properties of condensed matter (e.g. thermal expansion).

Arrhenius equation an equation of the form

$$k = A\exp\left(-\frac{E_a}{RT}\right)$$

where k is the rate constant of a given reaction, E_a is the (apparent) activation energy, and A is a constant for a given reaction, called the pre-exponential factor. The equation can be written in logarithmic form:

$$\ln k = \ln A - E_a / RT$$

Association a group (or cluster) of atoms that are held together by forces weaker than a normal chemical bond. For example, liquid selenium forms a high degree of association that leads to anomalously high viscosity values (cf. Table 17.9).

Atom the smallest and most basic unit of an element that can exist. Atoms consist of a small dense nucleus of protons and neutrons surrounded by moving electrons. The number of electrons equals the number of protons, so that the overall charge is zero.

According to modern theories, an electron in an atom is in a state described by the solution (i.e. wave functions) of a wave equation, or Schrödinger equation, used in quantum mechanics. This determines the probability that the electron may be located in a certain volume of space, called an orbital, about the nucleus of the atom. The state of each electron can be characterized by a set of four quantum numbers (cf. also **Pauli exclusion principle**). The four quantum numbers are (1) the principal quantum number, n, (2) the orbital (or azimuthal) quantum number, l, (3) the magnetic quantum number, m, and (4) the spin quantum number, s.

An exact calculation of the energies and other properties of the quantum states are only possible for the simplest atoms, but various approximate methods, which give useful results, have been proposed.

Atomic number the number of protons in the nucleus of an atom or the number of electrons revolving around the nucleus in a neutral atom.

Boiling point the temperature of a liquid at which visible evaporation occurs throughout the bulk of the liquid, and at which the vapour pressure of the liquid equals to external atmospheric pressure; the temperature remains constant until all the liquid has evaporated (i.e. at the boiling point, liquid and vapour can exist together in equilibrium at a given pressure). The variation of boiling point with pressure can be obtained from the Clausius–Clapeyron equation (see Subsection 1.2.2.1); boiling points are usually quoted for standard atmospheric pressure (101,325 Pa).

Brownian movement (or motion) the continuous, irregular motion in all directions of small particles about 1 μm in diameter when held in suspension in a liquid. It is a visible demonstration of molecular bombardment by the molecules of the liquid. The smaller the suspended particles are, the more noticeable the motion is.

Colloid a substance consisting of particles of ultramicroscopic size (1–500 nm; 10^3 to 10^9 atoms are contained). Colloidal particles exhibit Brownian movements.

***d*-block elements** the block of elements in the periodic table consisting of scandium, yttrium, lanthanum, and the elements of three main transition series (see **Transition elements**) together with Group IIB; their outer electron configurations are $(n-1)d^x ns^y$, where $x = 1 - 10$, $y = 1$ or 2(see Table 1.7). All main transition elements have similar thermophysical properties.

Degrees of freedom the number of independent variables, or parameters, needed to describe the configuration of a system. For example, in a monatomic gas, each atom may be allotted three degrees of freedom, corresponding to the three coordinates in space required to specify its position. The mean energy per atom for each degree of freedom is the same, according to the principle of equipartition of energy, and is equal to $kT/2$ for each degree of freedom (see **Internal energy of harmonic oscillators**).

Diagonal relationship a relationship within the periodic table by which certain elements in the second period have a close thermophysical similarity to their diagonal neighbours in the next group of the third period. This is particularly noticeable with the following pairs: lithium and magnesium, beryllium and aluminium, boron and silicon, etc. The reason for this relationship is a combination of the trends to increase size (i.e. atomic and ionic radii of elements) down a group and to decrease size along a period.

Electron configuration the way the electrons fill the various energy levels of the atom. The number of electrons in the outermost shell (i.e. the shell farthest away from the nucleus, by position) controls the thermophysical properties of an element. Elements with similar outer electron configurations behave in very similar ways.

Element a substance that cannot be decomposed into a simpler substance. The elements are the basic building blocks of all matter. The term 'element' is also used frequently in the meaning of 'simple substance or elementary substance'.

Energy the quantity that is the measure of the capacity of a body or a system for doing work. Energy is conventionally classified into two forms: kinetic energy and potential energy.

Kinetic energy is the energy of motion and is usually defined as the work that will be done by the body possessing the energy when it is brought to rest. In classical physics, it can be shown that a particle of mass m with speed v has translational kinetic energy, $mv^2/2$, while a rotating body with moment of inertia I about its axis of rotation and angular velocity ω has rotational kinetic energy $I\omega^2/2$.

Potential energy is energy stored in a body or system as a consequence of its position, shape, or state. For example, if a body of mass m is raised to a height h above the ground, the potential energy is mgh, where g is the acceleration of free fall (g is a fundamental constant). The potential energy includes electrical energy, nuclear energy, and chemical energy.

Internal energy is a thermodynamic characteristic function; the sum of the energies of the atomic (molecular) motions within a body (see **Internal energy** and **Internal energy of harmonic oscillators**).

Energy level we first go through our lesson on atomic structure. The Bohr model of the atom may still provide a simple, excellent basis for an understanding of atomic structure. According to the Bohr model, electrons travel around the nucleus of the atom in shells (or orbitals), which were described as imaginary spherical surfaces roughly concentric with the nucleus; each shell is associated with a particular energy level (i.e. electrons travel in specific shells), and that shells farther away from the nucleus are associated with higher energy levels.

The energy levels in atoms have been given the letter designations K, L, M, N, or 1, 2, 3, 4, ... Level K or 1 is the lowest-energy level. Each energy level can hold only a certain number of electrons. Lower-energy levels are usually filled before electrons are positioned in higher-energy levels. The following table gives the maximum number of electrons for each level, or shell.

Energy level†		Maximum number Of electrons††
Letter designation	Number (n)‡	
K	1	2
L	2	8
M	3	18
N	4	32
O	5	50
P	6	72
Q	7	98

† Traditionally, these levels, or the orbits corresponding to them, are referred to as shells.
‡ The principal quantum number.
†† The maximum number of electrons in a given shell is $2n^2$ (cf. **orbital** and **Pauli exclusion principle**).

Enthalpy (symbol H) a thermodynamic function of a system equal to the sum of its internal energy and displacement work (i.e. the product of its pressure, P, and volume V), namely,

$$H = U + PV$$

The total differential of the enthalpy is:

$$dH = dU + PdV + VdP$$

For a reversible process at constant pressure the work done by the system is equal to the product of pressure times the change of volume; the heat absorbed in such a process is equal to the increase in enthalpy of the system. The enthalpy is of importance for the description of processes proceeding at constant pressure.

Entropy (symbol S) a property of a system that changes, when the system undergoes a reversible change (or quasi-static change), by an amount equal to the energy δQ absorbed by the system divided by the thermodynamic temperature, i.e. $dS = \delta Q/T$. However, all real processes are, to a certain extent, irreversible changes; in any closed system, an irreversible change is always accompanied by an increase in entropy. (Closed system: a system that may exchange only energy with an outside environment, but cannot exchange matter.)

In a statistical mechanical sense, entropy can be thought of as a measure of the molecular disorder of a system; the higher the entropy, the greater the disorder. At absolute zero, the entropy is equal to zero (i.e. $S_{T=0} = 0$).

Equation of state equations that show the relationship between the pressure, volume, and temperature of a substance. The simplest is the ideal gas law:

$$PV = nRT$$

where n is the amount of substance (the SI unit of amount of substance is the mole), and R is the molar gas constant. The ideal gas is a simple model of a real gas, assuming that the particles are of negligible size, and have few mutual interactions; experimentally, the behaviour of a real gas departs more or less seriously from that of an ideal gas at high pressures or low temperatures. A more accurate equation of state (for one mole) is:

$$\left(P + \frac{a}{V^2}\right)(V - b) = RT$$

where a and b are material parameters, related to the effect of intermolecular forces and the size of molecules, respectively. This equation is well known as the van der Waals equation and is a very convenient way of describing the general behaviour of a real gas and the transition to the liquid state. If a and b are determined for one mole of gas, the van der Waals equation of state for n moles becomes

$$\left(P + n^2\frac{a}{V^2}\right)(V - nb) = nRT$$

Equipartition of energy the principle (or theory) that the mean energy of the molecules of a gas in a large sample under thermal equilibrium is equally divided among the various degrees of freedom of the molecules. The average energy of each degree of freedom is equal to $kT/2$. This principle is based on the Maxwell–Boltzmann distribution of classical statistical mechanics. The principle was later extended to the vibrations of atoms in crystals (see **Internal energy of harmonic oscillators**).

Evaporation the conversion of a substance (e.g. a metal) into a vapour at high temperatures, from the liquid state. For any liquid, the speed of evaporation depends on the temperature of the liquid. The evaporation process involves cooling of the liquid: at the surface of the liquid, some of the liquid atoms (molecules) with the highest kinetic energies escape into the gas phase, and consequently, there is a fall in the liquid's temperature.

f-block elements the block of elements in the periodic table consisting of the lanthanide series (from cerium to lutetium) and actinide series (from thorium to lawrencium). They are characterized by having two s-electrons in their outer (n) shell and f-electrons in their inner ($n-2$) shell. Thorium $(6d^2 7s^2)$ is an exceptional case.

Fluctuation a deviation of a physical quantity value x from its mean value $\bar{x}$ averaged over time: fluctuation $\delta \equiv x - \bar{x}$. For example, the pair distribution function $g(r)$ for any value of r represents the mean value, averaged over time, at that distance; large statistical fluctuations from this mean value will occur due to the thermal motions of the atoms (molecules, ions). From the microscopic point of view, the concept of fluctuation is of great importance for a clear understanding of the behaviour of metallic liquids.

Fundamental or universal constants the physical constants, or parameters, that do not change throughout the universe. For example, the speed of light, the charge of an electron or proton, the Planck constant, the Boltzmann constant, the Avogadro constant, the molar gas constant, the gravitational constant, the magnetic constant, the electric constant.

Glass by general agreement, it is said that a liquid on being cooled becomes a glass when the viscosity equals 10^{12} Pa s.

A glass has the disordered arrangement of the liquid from which it is derived by cooling below the freezing point, without crystallization, and has the elastic properties of an isotropic solid; a glass may be said to be a non-crystalline solid. Nevertheless, since the structures of glasses closely resemble those of liquids, glasses are often regarded as supercooled liquids.

Ground state the state of a system (e.g. atom) with the lowest stable energy. For example, in the case of the hydrogen, the single electron is moving around the nucleus in the shell closest to the nucleus (the shell associated with the lowest energy). In this situation, the electron is said to be in its lowest energy level and the atom is in its ground state. (As the electron moves, it neither gains nor loses energy; its energy of motion exactly counterbalances the attraction of the nucleus.)

When an electron is at an energy level above the lowest level, the atom is said to be in an *excited state*.

An electron can also absorb so much energy that it escapes completely from the atom. This process is called *ionization*. (Generally speaking, the process of removing one or several electrons from a neutral atom (or adding one or several electrons to a neutral atom), i.e. the process of forming ions is called *ionization*.)

Grüneisen relation a relation, or law, derived from the equation of state of solids. The Grüneisen relation can be expressed in the form:

$$\alpha = \frac{\gamma_G \kappa_T C_V}{V}$$

where α is the volume expansivity, κ_T is the isothermal compressibility, C_V is the constant-volume heat capacity, V is the volume, and γ_G is called the Grüneisen constant.

The Grüneisen relation is satisfied experimentally for cubic crystals, taking γ_G as independent of temperature. Generally speaking, however, every solid has its own Grüneisen constant; its value is between 1 and 3 for many crystalline solids (cf. Table 12.2), and is slightly dependent on the temperature at which the measurements are made.

Hard-sphere or rigid-sphere model the model which treats atoms (ions) as hard spheres (rigid spheres). On the basis of the hard-sphere model, expressed in terms of the packing fraction, several thermophysical properties of liquid metals can be explained quite successfully. The packing fraction η is defined as the fraction of the total volume occupied by the hard spheres, namely,

$$\eta = \frac{4}{3}\pi\left(\frac{\sigma}{2}\right)^3 \frac{N}{V} = \frac{\pi}{6} n_0 \sigma^3$$

where σ is the (effective) hard-sphere diameter, N is the total number of hard spheres in the volume V, and n_0 ($= N/V$) is the average number density. The packing fraction is determined by fitting a liquid structure factor, which is directly observed in X-ray or neutron experiments, and is about 0.46 for liquid metals at their melting point temperatures.

Heat a particular form of energy connected with the temperature increase of a substance. A supply of heat causes a temperature increase. The relation for the supply of heat and temperature increase is determined by the heat capacity (i.e. a material property). In a phase transition, however, the heat capacity tends to infinity (because the supply or release of heat does not lead to a change in temperature) and the definition mentioned above can no longer be applied.

Helmholtz free energy and Gibbs free energy a measure of a system's ability to work. The Helmholtz free energy, F, and the Gibbs free energy, G, are defined, as follows:

$$F = U - TS$$

and

$$G = U + PV - TS = H - TS$$

For an infinitesimal change,

$$dF = dU - TdS - SdT = (TdS - PdV) - TdS - SdT$$

$$\therefore\ dF = -SdT - PdV$$

At constant temperature ($dT = 0$),

$$dF = -PdV = dU - TdS$$

Thus the change dF is less than the change dU by the quantity TdS. This is the unavailable energy arising from entropy changes in the system; the energy that we can extract from the system is always less than the energy available from the internal energy.

Similarly, at constant temperature,

$$dG = dH - TdS$$

At constant temperature and constant pressure, the maximum work that the system can perform (dG) is the change in enthalpy H (= $U + PV$) less the unavailable energy TdS.

Inert-pair effect an effect seen especially in the Groups IIIA and IVA of the periodic table, in which the heavier elements in the group tend to form ions or compounds with a valence two lower than the expected group valence, e.g. Tl^{1+}, Pb^{2+}, etc. This behaviour of electrons is called the inert-pair effect. An explanation of the behaviour may be provided on the basis of the energy levels in atoms (or also ionization energies).

Internal energy (symbol *U*) the sum of the kinetic energies of the atomic (or molecular) motions and the potential energies of atomic (or molecular) interactions within a body. For a monatomic gas, the internal energy is almost entirely atomic kinetic energy. The internal energy is also called the thermal energy. In a closed system, the internal energy is identical to the total energy of the system. The change of the internal energy, ΔU, in any change of state is given by the sum of work, ΔW, and heat ΔQ, exchanged with the environment;

$$\text{internal energy} = \text{work} + \text{heat}: \Delta U = \Delta \mathrm{W} + \Delta Q$$

This statement, which is equivalent to the law of conservation of energy, is known as the first law of thermodynamics.

Internal energy of harmonic oscillators the energy ε of a harmonic oscillator of mass m can be expressed by six terms, as follows:

$$\varepsilon = \frac{1}{2}mv_x^2 + \frac{1}{2}mv_y^2 + \frac{1}{2}mv_z^2 + \frac{1}{2}Cx^2 + \frac{1}{2}Cy^2 + \frac{1}{2}Cz^2$$

where C is the constant of proportionality. According to the equipartition particle, the average energy of each degree of freedom is equal to $kT/2$, not only for kinetic energy but also for potential energy (the mathematical form of potential energy, e.g. $Cx^2/2$, is similar to that of kinetic energy, e.g. $mv_x^2/2$): for each of the six terms the average energy $\bar{\varepsilon}$ is equal to $kT/2$, so that $\bar{\varepsilon}$ value for the simple oscillator is

$$\bar{\varepsilon} = \frac{1}{2}kT \times 6 = 3kT$$

The internal energy U of N harmonic oscillators becomes

$$U = 3NkT$$

For a mole of substance, the internal energy, or the thermal energy, leads to

$$U = 3RT$$

Incidentally, molar heat capacity C_V at constant volume is

$$C_V = \left(\frac{\partial U}{\partial T}\right)_V = 3R \approx 25\ \text{Jmol}^{-1}\text{K}^{-1}$$

Ion an atom or group of atoms that has gained or lost one or more electrons, and therefore has a net positive or negative charge. A positive ion (or cation) contains less electrons than are necessary for the atom or group to be neutral, a negative ion (or anion) contains more.

Ionic bond a chemical bond in which one atom transfers one or more electrons to another atom. As a result, ions are held together by the attraction of opposite charges.

Ionic radius a value assigned to the radius of an ion in a crystalline solid, based on the assumption that ions are spherical with a definite size, that is, ions may be regarded as hard spheres; the distance at which the ions come into contact is the equilibrium interionic distance. X-ray diffraction can be used to measure the internuclear distance in crystalline solids. (The values for ionic radii of elements depend on the method of measurement.) In general, negative ions have larger ionic radii than positive ions. The larger the negative charge, the larger the ion; the larger the positive charge, the smaller the ion.

Ionization energy or ionization potential the minimum energy expended to remove an electron from a given atom, which is a stationary bound atomic state, to infinity (i.e. a distance that there is no electrostatic interaction between ion and electron). The farther away from the nucleus an electron is, the less it is attracted by the positive charge of the nucleus, and the less energy is needed to remove it away from the atom; the electron is always removed from the outermost energy level.

The energy to remove the least strongly bound electron is the first ionization energy (or potential). Second, third and higher ionization energies can also be measured.

Isotopes atoms of the same element that contain the same number of protons and electrons but different numbers of neutrons.

Laminar flow steady (or streamline) flow of a fluid in which the fluid moves in layers without turbulence, so that successive particles passing the same point have the same velocity. It occurs at low Reynolds numbers, i.e. low velocities, high (molecular) viscosities, low densities or small dimensions.

Lanthanoid contraction the $4f$ orbitals in the atoms of the lanthanoid series are not very effective in shielding the outer electrons from the nuclear charge. In going across the series, the increasing nuclear charge (i.e. increase in the number of protons) pulls hard on the electrons in the outer shells, which causes a contracting of the atomic and ionic radii (cf. Figure 1.11). This effect is known as the lanthanoid contraction.

Lanthanoids a series of elements in the periodic table, arranged in order of their atomic number (or proton number) from lanthanum ($_{57}$La) to lutetium ($_{71}$Lu) inclusive. The lanthanoids all have two outer s-electrons; in the lanthanoid series, an increasing atomic number corresponds to increase in number of $4f$ electrons. In fact, the $4f$ and $5d$ levels are close in energy and the filling is not smooth.

The lanthanoid metals commonly form M^{3+} ions (their main compounds contain M^{3+} ions). Cerium also has the highly oxidizing Ce^{4+} state, and europium and ytterbium have a M^{2+} state.

Latent heat the quantity of heat absorbed or released in an isothermal phase transition (e.g. from solid to liquid at the melting point or from liquid to gas at the boiling point).

In a phase transition proceeding at constant pressure and constant temperature, the latent heat received (in melting and boiling) or delivered (in solidifying and condensation) is equal to the change of enthalpy of a substance.

Logarithmic scale in physical sciences, a graph plotted on a logarithmic scale is often used. For example, consider the following equation:$y = x^n$, If a curve $y = x^n$ is plotted on graph paper with logarithmic scales on both axes, the result is a straight line of slope n, i.e.$\log y = n \log x$, which enables n to be determined (cf. Section 8.4).

Mass number or nucleon number the total number of nucleons in an atomic nucleus, i.e. mass number, or nucleon number, $A = Z + N$, where Z is the atomic number, or proton number, and N is the neutron number. (Nucleon: generic term for proton and neutron.)

Mean free path the mean distance of free flight of a particle (e.g. atom, molecule, electron in metals) between two successive collisions with other particles. Incidentally, the average time that elapses between their collisions is called the mean free time of flight.

Melting point the temperature at which a solid changes into a liquid. At the melting point temperature, the solid and liquid phases of a substance can exist in equilibrium at a defined pressure, normally standard pressure of 101,325 Pa. A pure substance under standard pressure has a single reproducible melting point, which is used as a material constant.

Metastable state a state of pseudo-equilibrium (e.g. supercooled liquid). It is often achieved by attaining the state (e.g. the liquid state) very slowly (cf. Figure 1.1(b)); a slight disturbance will produce the stable state.

Molar a term denoting that an extensive physical property is being expressed per amount of substance; in SI units, this means 'per mole'. For example, molar mass is the mass of one mole of a substance: molar mass $\left(\text{kg mol}^{-1}\right)$ = Avogadro constant $\left(\text{mol}^{-1}\right)$ × particle mass (kg).

Molar gas constant (symbol *R*) a fundamental constant which is the product of the Avogadro constant N_A and the Boltzmann constant k, i.e.$R = N_A k$.

According to the kinetic theory of gases, $RT = 2(mN_A\overline{v^2}/2)/3$, where $\overline{v^2}$ is the mean square velocity. This equation shows that R is equal to two-thirds of the total translational energy of molecules of one mole of a gas at a temperature of 1 K.

Mole the SI unit of amount of substance, defined as the amount of substance of a system that contains as many as elementary entities as there are atoms in 0.012 kg ^{12}C. The elementary entities must be specified (e.g. atoms, molecules, ions, electrons). 1 mole is equal to 6.022×10^{23} $\left(\text{mol}^{-1}\right)$, i.e. Avogadro constant (mol^{-1}) (see Appendix 6); it is simply a number of entities, or items.

Orbital the orbital represents a region of space in which the electron can be found with a high probability. Generally speaking, a state defined by quantum numbers n and l is called an orbital.

Pair distribution function $g(r)$ $g(r)$ is proportional to the probability of finding another atom, at a distance r from the reference atom located at $r = 0$. Incidentally, the function $4\pi r^2 n_0 g(r)$ is generally called the radial distribution function.

The structures and properties of liquids can be expressed in terms of $g(r)$.

Pauli exclusion principle the principle (applying to Fermi particles or fermions) that no two identical particles (e.g. electrons) in an atom can have all four quantum numbers the same: in other words, each energy level can hold only two electrons (Fermi particles) and these must have opposite spins (spin: the intrinsic angular momentum of an elementary particle.)

The Pauli exclusion principle gives an understanding of the structure of the electron shell of atoms.

***p*-block elements** the block of elements in the periodic table consisting of Groups IIIA through VIIIA. These are representative elements, and are composed of metals, metalloids (semimetals and semiconductors), and non-metals. The outer electron configurations of these elements all have the form $ns^2\ np^x$, where $x = 1 - 6$.

Periodic law the principle that the physical and chemical (or thermophysical) properties of elements are a periodic function of their atomic number, or proton number.

Periodic table the classification of chemical elements, in tabular form, in order of their atomic numbers (or proton numbers). The elements fall into vertical columns, known as groups (or families). Horizontal rows in the table are periods; in each period, the elements are arranged in order of their atomic number.

There are certain general features of thermophysical behaviour shown in the periodic table. In moving down a group, there is an increase in metallic character because of the increased size of the atom. However, elements in the same group have the same valence electron configuration; as a result, their thermophysical behaviour will be similar. In going across a period, there is a change from metallic behaviour to non-metallic, because of the increasing number of electrons in the outermost shell. However, the transition metals fill their inner electron shells after outer electron shells, so that they show similar thermophysical behaviour. Overall, metallic elements tend to be those on the left and towards the bottom of the table; non-metallic elements are towards the top and the right.

There is also a significant difference between the elements of the second short period (e.g. lithium, beryllium) and the other elements in their respective groups (see **Diagonal relationship**).

Quantum (plural, quanta) the smallest amount of energy that a system can gain or lose; the energy may be exchanged only in definite amounts. The change in energy corresponding to a quantum is very small and is only noticeable on an atomic scale. This concept forms the basis of quantum mechanics.

Radioactive element an element, or a radioactive element, that is emitting radiation is said to be undergoing radioactive decay, or nuclear decay. Some elements do so spontaneously (e.g. $^{226}_{88}Ra$). When a radioactive element undergoes nuclear decay, it emits radiation and becomes a different element; this is called a nuclear transformation or transmutation. The ability or property of a substance to emit penetrating rays (or radiation) from its nucleus is defined as radioactivity.

There are three types of radiation: alpha particles, beta particles, and gamma rays. Alpha particles (α particles) have a mass number of 4 and a charge of +2, and little penetrating ability (i.e. alpha particles are helium nuclei, $^{4}_{2}He$); they can be stopped by a thin sheet of paper. Beta particles (β particles) consist of electrons (e^-) or positrons (e^+), and have only slight penetrating power; they can be stopped by a thin sheet of almost any metal. Gamma rays (γ rays) have neither mass nor charge; γ rays are made up of very high energy (strong penetrating ability), and travel straight through a magnetic field.

Rare earth elements the lanthanoids, plus scandium and yttrium, are known generically as rare earth elements (cf. Table 10.4) because of the thermophysical similarities of these 17 elements.

Relative atomic mass the average mass per atom of a given specimen of an element, expressed in unified atomic mass units. The value depends on the isotopes present in the specimen. It was formerly called atomic weight. Incidentally, the (unified) atomic mass unit is a unit of mass equal to 1/12 of the mass of an atom of carbon-12 (^{12}C).

Relative atomic mass corresponds to numerical value, i.e. physical quantity = numerical value × unit.

Relative density the density of a substance divided by the density of water. This quantity was formerly called specific gravity.

***s*-block elements** the elements of the first two main groups of the periodic table, i.e. Groups IA and IIA metals. The outer electron configurations of these metals all have inert-gas structures plus outer ns^1 (IA) or ns^2 (IIA) electrons; the valence electrons in the metals are only *s*-electrons. Typically, the *s*-block elements are reactive metals forming stable ionic compounds containing M^{1+} and M^{2+}.

Shell or electron shell according to quantum physics, the electrons in an atom are restricted to certain orbitals, or energy levels. These orbitals, or energy levels, are often referred to as shells (cf. **Energy level** and **Orbital**).

State of matter one of three physical states in which matter can exist, i.e. solid, liquid, or gas. By applying energy, matter may be converted from a solid state to a liquid or gaseous state, and a liquid may turn into a gaseous state. Plasma is sometimes regarded as the fourth state of matter.

Supercooling the cooling of a liquid to below its freezing point without a change from the liquid to solid state taking place. A supercooled liquid is a metastable state, and the introduction of the smallest quantity of the solid, or a small mechanical disturbance will start solidification. The analogous cooling of a vapour to make it supersaturated.

Transition elements generally speaking, a set of elements in the periodic table in which filling of electrons in an inner *d* or *f*-level occurs (see Table 1.7).

The classification of chemical elements is valuable only in so far as it shows similar thermophysical behaviour (e.g. Figures 10.9 and 14.6 (surface tension) and Figures 10.10 and 15.7 (viscosity)). Thus, from the standpoint of materials process science, the authors, in this book, term the following elements of the three groups of transition metals as 'transition elements'.

Ti	V	Cr	Mn	Fe	Co	Ni	Cu
Zr	Nb	Mo	Tc	Ru	Rh	Pd	Ag
Hf	Ta	W	Re	Os	Ir	Pt	Au

These elements are usually called three 'main transition series'. They have outer electron configurations of the form $(n-1)\,d^{2\text{–}10}ns^{0\text{–}2}$. In the transition metals, electrons fill inner shells $(n-1)$ after outer shells n, and there is additional binding from inner electron shells; several of an element's characteristic properties (e.g. high values of surface tension, viscosity, and melting and boiling points) arise from high binding energy.

Triple point the temperature and pressure at which the vapour (gas), liquid, and solid phases of a substance can be together in equilibrium.

The temperature of the triple point for pure water is defined to be 273.16 K (610.6 Pa). This is the fixed point for the thermodynamic temperature scale.

Valence electrons the electrons that occupy the outermost shell of an atom. They often determine the chemical and physical behaviour of the element.

Vapour pressure the pressure exerted by a vapour. For a vapour in equilibrium with its liquid (if the liquid is in a closed space, the vapour pressure will reach an equilibrium value), the vapour pressure depends on the temperature of the liquid and is known as the saturated vapour pressure at that temperature. Every liquid has its own characteristic equilibrium vapour pressure.

X-rays energetic electromagnetic waves (or radiation) of short wavelength; the range of wavelength is 10^{-11} to 10^{-9} m. Atoms of all elements emit a characteristic X-ray spectrum when they are bombarded by electrons. X-rays have a significant penetration depth in materials due to their high energy. X-rays can be reflected and refracted by suitable materials; they also show interference and diffraction effects. X-rays are therefore used for measurement of thickness, materials testing, examination of internal structure, and quality control.

GENERAL REFERENCES

1. A. Isaacs ed., *A Concise Dictionary of Physics*, Oxford University Press, 1990.
2. V. Illingworth ed., *The Penguin Dictionary of Physics*, 2nd ed., Penguin Books, 1991.
3. J. Daintith ed., *A Concise Dictionary of Chemistry*, Oxford University Press, 1990.

Appendix 1
Periodic Table of the Elements

Period	IA	IIA	IIIB	IVB	VB	VIB	VIIB	VIIIB			IB	IIB	IIIA	IVA	VA	VIA	VIIA	VIIIA
Group	1	2	3	4	5	6	7	8	9	10	11	12	13	14	15	16	17	18
1	Non-metal $_{1}$H																	Non-metal $_{2}$He
2	Metal $_{3}$Li	Metal $_{4}$Be											Metalloid* $_{5}$B	Non-metal $_{6}$C	Non-metal $_{7}$N	Non-metal $_{8}$O	Non-metal $_{9}$F	Non-metal $_{10}$Ne
3	Metal $_{11}$Na	Metal $_{12}$Mg	Transition elements										Metal $_{13}$Al	Metalloid $_{14}$Si	Non-metal $_{15}$P	Non-metal $_{16}$S	Non-metal $_{17}$Cl	Non-metal $_{18}$Ar
4	Metal $_{19}$K	Metal $_{20}$Ca	Metal $_{21}$Sc	Metal $_{22}$Ti	Metal $_{23}$V	Metal $_{24}$Cr	Metal $_{25}$Mn	Metal $_{26}$Fe	Metal $_{27}$Co	Metal $_{28}$Ni	Metal $_{29}$Cu	Metal $_{30}$Zn	Metal $_{31}$Ga	Metalloid $_{32}$Ge	Metalloid $_{33}$As	Metalloid $_{34}$Se	Non-metal $_{35}$Br	Non-metal $_{36}$Kr
5	Metal $_{37}$Rb	Metal $_{38}$Sr	Metal $_{39}$Y	Metal $_{40}$Zr	Metal $_{41}$Nb	Metal $_{42}$Mo	Metal $_{43}$Tc	Metal $_{44}$Ru	Metal $_{45}$Rh	Metal $_{46}$Pd	Metal $_{47}$Ag	Metal $_{48}$Cd	Metal $_{49}$In	Metal $_{50}$Sn	Metal $_{51}$Sb	Metalloid $_{52}$Te	Non-metal $_{53}$I	Non-metal $_{54}$Xe
6	Metal $_{55}$Cs	Metal $_{56}$Ba	Metal $_{57}$La †	Metal $_{72}$Hf	Metal $_{73}$Ta	Metal $_{74}$W	Metal $_{75}$Re	Metal $_{76}$Os	Metal $_{77}$Ir	Metal $_{78}$Pt	Metal $_{79}$Au	Metal $_{80}$Hg	Metal $_{81}$Tl	Metal $_{82}$Pb	Metal $_{83}$Bi	Metal $_{84}$Po	Metalloid $_{85}$At	Non-metal $_{86}$Rn
7	Metal $_{87}$Fr	Metal $_{88}$Ra	Metal $_{89}$Ac ‡															

† Lanthanides	Metal $_{58}$Ce	Metal $_{59}$Pr	Metal $_{60}$Nd	Metal $_{61}$Pm	Metal $_{62}$Sm	Metal $_{63}$Eu	Metal $_{64}$Gd	Metal $_{65}$Tb	Metal $_{66}$Dy	Metal $_{67}$Ho	Metal $_{68}$Er	Metal $_{69}$Tm	Metal $_{70}$Yb	Metal $_{71}$Lu
‡ Actinides	Metal $_{90}$Th	Metal $_{91}$Pa	Metal $_{92}$U	Metal $_{93}$Np	Metal $_{94}$Pu	Metal $_{95}$Am	Metal $_{96}$Cm	Metal $_{97}$Bk	Metal $_{98}$Cf	Metal $_{99}$Es	Metal $_{100}$Fm	Metal $_{101}$Md	Metal $_{102}$No	Metal $_{103}$Lr

* In this periodic table, the elements are classified into three groups, (or classes): metals, metalloids, and non-metals. The metalloids fall halfway between metals and non-metals. Metalloids have some properties like those of metals and other properties like those of non-metals. While the classification is not clear cut, typical metalloids are boron, silicon, germanium, arsenic, and tellurium. They are electrical semiconductors and their oxides are amphoteric.

Appendix 2
Derivation of the Linear Equation Satisfying the Condition of the Minimum *S* Value (i): a Slope = *m*, an Intercept = 0

Suppose the linear equation satisfying the condition of the minimum relative standard deviation S for a given set of data (x_i, y_i) is written as

$$Y_i = mx_i \tag{A2.1}$$

where m is the slope of the line $(x_i > 0,\ y_i > 0,\ m > 0)$. Assuming the following expression:

$$L = \sum_{i=1}^{n} \left(\frac{y_i - Y_i}{Y_i} \right)^2 \tag{A2.2}$$

substituting the value of Y_i from Eq. (A2.1) into Eq. (A2.2), we have for L

$$L = \sum \left(\frac{y_i}{mx_i} - 1 \right)^2 \tag{A2.3}$$

Differentiating Eq. (A2.3) with respect to m, we obtain

$$\frac{dL}{dm} = -\frac{2}{m^3} \sum \left(\frac{y_i}{x_i} \right)^2 + \frac{2}{m^2} \sum \left(\frac{y_i}{x_i} \right) \tag{A2.4}$$

For relative standard deviation S to be a minimum, the following condition must be satisfied:

$$\frac{dL}{dm} = 0 \tag{A2.5}$$

Applying the above condition, the final result obtained is

$$m = \frac{\sum \left(\frac{y_i}{x_i} \right)^2}{\sum \left(\frac{y_i}{x_i} \right)}, \quad \text{or} \quad m = \frac{\overline{\left(\frac{y_i}{x_i} \right)^2}}{\overline{\left(\frac{y_i}{x_i} \right)}} \tag{A2.6}$$

Appendix 3
Derivation of the Linear Equation Satisfying the Condition of the Minimum *S* Value (ii): a Slope = *a*, an Intercept = *b*

Suppose the linear equation satisfying the condition of the minimum relative standard deviation S for a given set of data (x_i, y_i) is written as

$$Y_i = ax_i + b \tag{A3.1}$$

where a is the slope and b is the intercept on the vertical axis ($x_i > 0$, $y_i > 0$, $a > 0$, $(ax + b) > 0$). Assuming the following expression:

$$L = \sum_{i=1}^{n} \left(\frac{y_i - Y_i}{Y_i} \right)^2 \tag{A3.2}$$

Substituting the value of Y_i from Eq. (A3.1) into Eq. (A3.2), we have for L

$$L = \sum \left(\frac{y_i}{ax_i + b} - 1 \right)^2 \tag{A3.3}$$

Differentiating Eq. (A3.3) with respect to a, we obtain

$$\frac{\partial L}{\partial a} = -2 \sum \frac{x_i y_i^2}{(ax_i + b)^3} + 2 \sum \frac{x_i y_i}{(ax_i + b)^2} \tag{A3.4}$$

Similarly, differentiating Eq. (A3.3) with respect to b, we obtain

$$\frac{\partial L}{\partial b} = -2 \sum \frac{y_i^2}{(ax_i + b)^3} + 2 \sum \frac{y_i}{(ax_i + b)^2} \tag{A3.5}$$

For relative standard deviation S to be a minimum, the following conditions must be satisfied:

$$\frac{\partial L}{\partial a} = -2 \sum \frac{x_i y_i^2}{(ax_i + b)^3} + 2 \sum \frac{x_i y_i}{(ax_i + b)^2} = 0 \tag{A3.6}$$

$$\frac{\partial L}{\partial b} = -2 \sum \frac{y_i^2}{(ax_i + b)^3} + 2 \sum \frac{y_i}{(ax_i + b)^2} = 0 \tag{A3.7}$$

Solving the above simultaneous equations for a and b on a computer, a practically exact solution is obtained.

Appendix 4
The SI (Systemme Internationale) Units

Physical quantity	Unit name	Abbreviation (symbol)
Length	metre	m
Mass	kilogram	kg
Time	second	s
Electric current	ampere	A
Thermodynamic temperature	kelvin	K
Amount of substance	mole	mol
Luminous intensity	candela	cd

Prefixes used with SI units†

Value	Prefix	Abbreviation
10^{-1}	deci	d
10^{-2}	centi	c
10^{-3}	milli	m
10^{-6}	micro	μ
10^{-9}	nano	n
10^{-12}	pico	p
10^{-15}	femto	f
10^{-18}	atto	a
10^{-21}	zepto	z
10^{-24}	yocto	y
10^{1}	deca	da
10^{2}	hecto	h
10^{3}	kilo	k
10^{6}	mega	M
10^{9}	giga	G
10^{12}	tera	T
10^{15}	peta	P
10^{18}	exa	E
10^{21}	zetta	Z
10^{24}	yotta	Y

† Multiples or decimal fractions of the basic SI units are designated by prefixes.

Appendix 5
Units and Unit Conversions

Quantity	Non-SI unit	SI unit	Conversion to SI units
Length	Å	m	$1\ \text{Å} = 10^{-10}\ \text{m} = 0.1\ \text{nm}$
Volume	l	m^3	$1\ \text{l} = 10^{-3}\ \text{m}^3 = 1\ \text{dm}^3$, $1\ \text{ml} = 10^{-6}\ \text{m}^3 = 1\ \text{cm}^3$
Frequency		$\text{Hz}(\text{s}^{-1})$	
Density	g cm^{-3}	kg m^{-3}	$1\ \text{g cm}^{-3} = 10^3\ \text{kg m}^{-3} = 1\ \text{Mg m}^{-3}$
Force	dyn (g cm s^{-2})	N (kg m s^{-2})	$1\ \text{dyn} = 10^{-5}\text{N} = 10\ \mu\text{N}$
Pressure	atm	Pa (N m^{-2})	$1\ \text{atm} = 101.325\ \text{kPa}$
	bar		$1\ \text{bar} = 10^5\ \text{Pa} = 0.1\ \text{MPa}$
	Torr(= mmHg)		$1\ \text{Torr} = (101,325/760) \approx 133.322\ \text{Pa}$
Energy, work	erg	J(N m = W s)	$1\ \text{erg} = 10^{-7}\ \text{J} = 0.1\ \mu\text{J}$
	eV		$1\ \text{eV} = 1.60218 \times 10^{-19}\ \text{J}$
	cal mol^{-1}	J mol^{-1}	$1\ \text{cal mol}^{-1} = 4.184\ \text{J mol}^{-1}$
Power	erg s^{-1}	$\text{W}(\text{J s}^{-1} = \text{V A})$	

Quantity	Non-SI unit	SI unit	Conversion to SI units
Surface tension	dyn cm^{-1}	N m^{-1}	1 dyn $cm^{-1} = 10^{-3}$ N m^{-1} = 1 mNm^{-1}
Surface energy	erg cm^{-2}	J m^{-2}	1 erg $cm^{-2} = 10^{-3}$ J m^{-2} = 1 mJ m^{-2}
Dynamic viscosity	P (g cm^{-1} s^{-1})	Pa s	1 P = 0.1 Pa s, 1 cP = 1 mPa s
Kinematic viscosity	St (cm^2 s^{-1})	m^2 s^{-1}	1 St = 10^{-4} m^2 s^{-1} = 1 cm^2 s^{-1}
Diffusivity	cm^2 s^{-1}	m^2 s^{-1}	1 cm^2 $s^{-1} = 10^{-4}$ m^2 s^{-1}
Electrical resistivity	Ω cm	Ω m	1 Ω cm = 10^{-2} Ω m
Temperature	°C	K	$t/°C = (T/K) - 273.15$
Heat capacity	cal deg^{-1}	J K^{-1}	1 cal deg^{-1} = 4.184 J K^{-1}
Thermal conductivity	cal cm^{-1} s^{-1} deg^{-1}	W m^{-1} K^{-1}	1 cal cm^{-1} s^{-1} deg^{-1} = 0.4184 kJ s^{-1} m^{-1} K^{-1}
Entropy		J K^{-1}	
Molar entropy		J mol^{-1} K^{-1}	
Internal energy		J	
Free energy		J	
Enthalpy		J	
Particle number		1	
Number density (particle)		m^{-3}	

Appendix 6
Fundamental Physical Constants in SI Units

Quantity or constant	Symbol	Value with units[†]
Speed of light in vacuum	c	$299{,}792{,}458\ \mathrm{m\ s^{-1}}$
		exact by definition
Charge of electron or proton; Elementary charge	e	$\pm 1.602176565 \times 10^{-19}\ \mathrm{C}$
Planck constant	h	$6.62606957 \times 10^{-34}\ \mathrm{J\,s}$
Avogadro constant	N_{A}	$6.02214129 \times 10^{23}\ \mathrm{mol^{-1}}$
Farady constant	F	$96{,}485.3365\ \mathrm{C\,mol^{-1}}$
Electron (rest) mass	m_{e}	$9.10938291 \times 10^{-31}\ \mathrm{kg}$
Molar gas constant	R	$8.3144621\ \mathrm{J\,mol^{-1}\,K^{-1}}$
Boltzmann constant $\left(R\,N_{\mathrm{A}}^{-1}\right)$	k	$1.3806488 \times 10^{-23}\ \mathrm{J\,K^{-1}}$
Standard acceleration of gravity	g	$9.80665\ \mathrm{m\,s^{-2}}$

[†]The numerical values are taken from 2010 CODATA.
(CODATA = Committee on Data for Science and Technology.)

Appendix 7
The Greek Alphabet

Upper case	Lower case	Name
Α	α	alpha
Β	β	beta
Γ	γ	gamma
Δ	δ	delta
Ε	ε	epsilon
Ζ	ζ	zeta
Η	η	eta
Θ	θ	theta
Ι	ι	iota
Κ	κ	kappa
Λ	λ	lambda
Μ	μ	mu
Ν	ν	nu
Ξ	ξ	xi
Ο	ο	omicron
Π	π	pi
Ρ	ρ	rho
Σ	σ	sigma
Τ	τ	tau
Υ	υ	upsilon

continued

(continued)

Upper case	Lower case	Name
Φ	φ,ϕ	phi
Χ	χ	chi
Ψ	ψ	psi
Ω	ω	omega

Appendix 8
Occam's Razor (or Ockham's Razor)

Occam's razor states that 'entities should not be multiplied beyond necessity', 'given two equally predictive theories, choose the simplest', or 'of two competing theories or explanations, all other things being equal, the simpler one is to be preferred'.

Incidentally, a quote attributed to Einstein states that 'theories should be as simple as possible, but no simpler'.

Appendix 9(a)
Calculated Values of Isothermal Compressibility and Structure Factor

Values of isothermal compressibility κ_T and structure factor S(0) calculated from Eq. (5.38b) or Eq. (5.39b), and Eq. (5.5b) (see Chapter 5), respectively, for some liquid metallic elements at their melting point temperatures, together with values of the physical quantities or parameters used for the calculations.†

Element		κ_T 10^{-11} m^2 N^{-1}	$S(0)$ 10^{-3}	γ_h	V 10^{-6} m^3 mol^{-1}	T_m K	$\Delta_l^g H_b$ 10^3 J mol^{-1}	ξ_T	ξ_E
Bismuth	Bi	4.25	9.26	1.15	20.79	544.6	152	1.468	1.849
Cadmium	Cd	3.06	10.8	1.23	14.0	594.2	100	1.346	2.813
Caesium	Cs	65.2	22.6	1.16	72.2	301.6	65.9	0.606	0.974
Gallium	Ga	2.14	4.73	1.08	11.4	302.9	254	2.702	1.133
Indium	In	2.95	6.47	1.12	16.3	429.7	226	2.045	1.367
Lead	Pb	3.39	8.72	1.20	19.404	600.6	180	1.627	1.909
Potassium	K	38.1	22.5	1.11	47.3	336.5	77.4	0.582	0.889
Rubidium	Rb	49.6	22.3	1.15	57.7	312.5	75.2	0.609	0.889
Sodium	Na	18.9	23.5	1.12	24.8	370.9	89.1	0.562	0.823
Thallium	Tl	3.92	10.4	1.21	18.01	577	162	1.371	1.717
Tin	Sn	2.62	6.47	1.11	17.0	505.1	291	2.029	1.238
Zinc	Zn	2.34	13.6	1.25	9.945	692.7	115	1.090	2.309

† Data, except for those on γ_h and ξ, are given in Chapter 17. Data on γ_h are taken from Kleppa (see refrence [2] in Chapter 12).
Data on ξ are listed in Table 12.1.

Appendix 9(b)
Values of $\gamma\kappa_T$

Values of $\gamma\kappa_T$ for some liquid metallic elements at their melting point temperatures.[†]

Element		γ[‡] $N\,m^{-1}$	κ_T $10^{-11}\,m^2\,N^{-1}$	$\gamma\kappa_T$ $10^{-10}\,m$
Bismuth	Bi	0.382	4.25	0.162
Cadmium	Cd	0.637	3.06	0.195
Caesium	Cs	0.069	65.2	0.450
Gallium	Ga	0.724	2.14	0.155
Indium	In	0.560	2.95	0.165
Lead	Pb	0.457	3.39	0.155
Potassium	K	0.110	38.1	0.419
Rubidium	Rb	0.0896	49.6	0.444
Sodium	Na	0.197	18.9	0.372
Thallium	Tl	0.459	3.92	0.180
Tin	Sn	0.555	2.62	0.145
Zinc	Zn	0.789	2.34	0.185

[†] Cf. Table 6.1.
[‡] Data are given in Chapter 17.

Appendix 10
Calculated Values of γ_h

Calculated values for the ratio of heat capacity $\gamma_h (\equiv C_P / C_V)$ for some liquid metallic elements at their melting point temperatures.

Metallic element		γ_h	
		Eq. (12.3)	Kleppa†
Bismuth	Bi	1.15	1.15
Cadmium	Cd	1.28	1.23
Caesium	Cs	1.17	1.16
Gallium	Ga	1.04	1.08
Indium	In	1.07	1.12
Lead	Pb	1.15	1.20
Potassium	K	1.13	1.11
Rubidium	Rb	1.13	1.15
Sodium	Na	1.11	1.12
Thallium	Tl	1.14	1.21
Tin	Sn	1.06	1.11
Zinc	Zn	1.28	1.25

† See reference [2] in Chapter 12.

Index

C

D

E

I

K

L

M